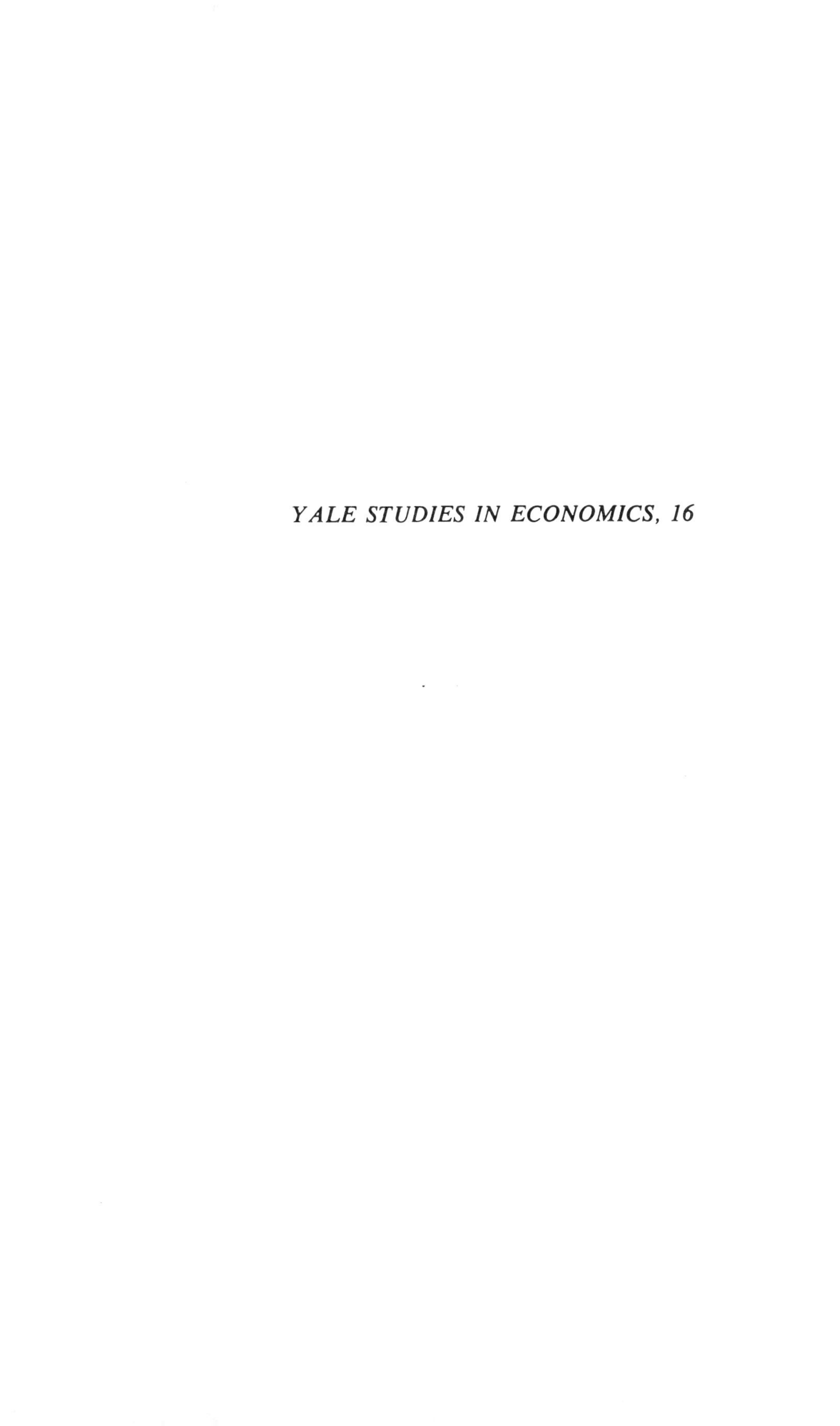

YALE STUDIES IN ECONOMICS, 16

STEEL DECISIONS AND THE NATIONAL ECONOMY

Henry W. Broude

YALE UNIVERSITY PRESS
NEW HAVEN AND LONDON, 1963

Designed by Wayne Peterson,
set in Times Roman type,
and printed in the United States of America by
The Carl Purington Rollins Printing-Office
of the Yale University Press.

Library of Congress catalog card number: 63–13958

ACKNOWLEDGMENTS

I wish to express my gratitude to Professor Wassily Leontief for his counsel and guidance during the preliminary writing of parts of this volume, and also to Professors Carl Kaysen and James S. Duesenberry, who commented on the development of several points. Professors William Fellner, Kent T. Healy, Arthur M. Okun, Richard Ruggles, and James Tobin all have made important contributions as to ideas and methods of approach. Dean John Perry Miller helped me in the formulation of some of the material and offered many penetrating criticisms.

I also wish to express my appreciation to the officers of the various steel companies who shared their insights and opinions with me. For their efforts in connection with the presentation of the materials in this study, I wish to thank Professors Louis H. Pollak and Harry H. Wellington, as well as Mrs. Alice S. Miskimin. I also wish to thank *The Quarterly Journal of Economics* for permission to include material which appeared in my article, "Bottleneck Phenomena and Cyclical Change: The Role of the Iron and Steel Industry," in that journal, Vol. 68, August 1954.

To Professors Lloyd G. Reynolds and John E. Sawyer, I am grateful for encouragement to develop this material, and also for their acting on my behalf to secure financial assistance through the grant to the Economics Department of Yale University from the Ford Foundation to aid faculty research, and through the grant of the Carnegie Corporation of New York to aid research in economic history. Neither the organizations nor the individuals mentioned in the foregoing, of course, bear any responsibility for ideas expressed or errors committed in this volume; they are all mine. They do, however, have my sincere appreciation for all they have done.

H.W.B.

CONTENTS

TABLES

FIGURES

STEEL DECISIONS AND
THE NATIONAL ECONOMY

ONE AN INTRODUCTION TO THE PROBLEM

Immediately after World War II, a controversy developed between government "experts" and representatives of the American iron and steel industry. At issue was the question whether industry was contemplating an expansion of its capacity sufficient to accommodate the needs of a growing, full-employment economy. At the time, this controversy accomplished little, for two very good reasons. First, there was no reassuringly reliable method for determining *adequate* iron and steel capacity relative to national economic needs. Second, there was little or no effective basis for communication between the government experts on the one hand and the industry representatives on the other, for the basic perspectives of each group were very different indeed.

The picture has changed in the fourteen years since this controversy reached its height, when industry adamantly insisted it was expanding enough and government vigorously insisted it was not. Changing attitudes on the part of the industry, changing personnel in government, and changing operational policies, as well as developments in technology and shifts in the structure of the economy, all have contributed to this. Furthermore, discrepancies that arose from a basic inability to acquire quantitative data have in many respects been modified as more sophisticated analytical techniques have appeared and been adopted. Indeed, the most recent developments in the industry raise the question whether the steel industry may not have to pay more attention to the surplus rather than the deficit capacity problem. Whether or not this is so, however, it has been appropriate, and it remains appropriate, to ask: what is adequate capacity?

Moreover, the controversy of the late '40s between the government and the iron and steel industry is a prototype of similar controversies in other areas in the economy. Atomic energy is an example:

> The Atomic Energy Commission called today for new proposals from private and public interests on the construction of nuclear power plants. . . . The Commission coupled the invitation with a statement that if the industry did not respond with acceptable proposals the Commission would ask Congress for money to construct the reactors.[1]

Another example is the reluctance of a large electrical manufacturer to embark upon the production of needed defense equipment because the enterprise looked "unprofitable." Both examples—and there are many more—suggest a continuing problem: how to resolve the gap that so often exists between "public" appraisal of need and private response thereto. The problem is especially severe in industries in which heavy capital outlay must precede the output of new products.

This study is concerned, in part at least, with a particular set of circumstances, which may be described as follows: (1) an economy unable, in the short run, to move away from its dependency on a basic input material, steel, even though price and supply factors would suggest such a movement, (2) an industry, iron and steel, unwilling to expand capacity as fast as national economic policy dictates, and (3) the special reasons for both of these phenomena.

Beyond treating this special case, there are broader objectives for this study. In the first place, we want to answer two

1. "U.S. Acts to Speed Atom Power Use," *The New York Times* (January 7, 1957), p. 1.

fundamental questions concerning the American iron and steel industry: (1) is the industry so strategically "placed" in the structure of the national economy that actions within it have pivotal significance to the well-being of the general economy? and (2) what policies, if any, can and should be instituted to affect behavior and performance in the industry?

As with so many other problems that touch on capital formation and the capital goods sector, the time dimension peculiarly determines what deserves study and what conclusions are likely to emerge. For example, is it reasonable to assume that (within limits) substitution will take place in our economy, and that, accordingly, no particular good can be said to be indispensable? To define within what range the economy is vitally dependent on a particular industry's output, we must determine whether substitution is likely to occur away from its product, within what time span this might conceivably take place, and at what cost.

There will, of course, be some circumstances in which costs are so formidable as to inhibit substitution, at least for a period of time long enough to permit the industry to impose the consequences of its actions on the economy. Such circumstances suggest that there may be times when the economy is unable to get out from under the industry's strategic structural position *in time* or *at costs* that will allow it to ignore what may be taking place within the industry. And though much of the following discussion deals with the potential bottleneck that the industry may represent, a condition of persistent excess capacity also raises questions of social cost that must be analyzed.

In connection with social costs, a few further remarks should be made. Longer-run trends in steel industry capacity and output may show the emergence of new uses for the metal along with a decline in present uses. Moreover, different patterns of use for the metal can be expected under different industry pricing policies. Price rigidity (because of the industry's oligopolistic structure) can be expected to accelerate the shift away from steel when substitution is possible, insofar as it manifests itself on the high-price side. But it is altogether possible that new special uses will shelter the industry from this trend, in spite of price rigidity. Another question arises: will the industry be hurt even in the absence of new demand? It does not necessarily follow that the industry will suffer in the face of a falling share of the market for basic metals, for it is rather to be expected that output and price decisions will follow a quite rational profit-maximizing pattern for oligopolists and that the industry will initiate price reductions to forestall substitution away from its product *only* insofar as this would increase net profits. Misallocation would occur, however, under these conditions; that is, less steel produced at prices higher than the "competitive" equilibrium would deprive the economy of potential output. This is a type of social cost to be weighed against other factors considered below.

If we assume for the moment that it is desirable to avoid the consequences of oligopoly, we may be led to a series of policy decisions which would enforce price flexibility on the industry. This suggests either (a) an enforced break-up of concentrated units of production and a restoration of more competitive conditions in the industry; (b) an enlarged program of governmental coercion (e.g., Congressional committee pressure) which would result in a more "elastic" pricing pattern, but which would leave the structure of the industry intact; (c) some form of outright government competition with the industry through the construction of alternative sources of supply at lower prices; or (d) a federally regulated price and output control mechanism similar in function to that of a public utilities commission.

The problem has, however, still another dimension—the short-run potential bottleneck condition. The society may, at times, face a threat to its steady advance or to its goal of general economic stability if its rate of growth is inhibited by reaching, in the short run, the ceiling of its steel-making capacity. It must be emphasized that this is recognized as a short-run contingency.

Even in the face of substitution possibilities, it is unreasonable to expect (and historical evidence contradicts the suggestion) that the economy, suddenly encountering forces pointing to a spurt in growth (either from military or nonmilitary causes), can quickly reconcile itself to a shortage in steel by shifting to substitute materials. Accordingly, in such circumstances, the bottleneck problem is a real one. Given that economic development does proceed in spurts, and given the basic erratic pattern of so-called exogenous disturbances to the ideal orderly progress of economic advance, we must ask, how much is at stake if a bottleneck of this sort arises? Here is a second type of social cost, i.e., the cost to stability and growth of sudden resource constraint, so abrupt that we have insufficient time to pull away and turn to alternative supplies of materials.

If we speculate for the moment on possible types of policy to avoid this cost, we face the question of maintaining slack in the industry. In effect, we are saying that the economy has a stake in considering as a "normal" operating rate, some output level short of 100 per cent. This might well be assured if the oligopoly structure of the industry were maintained and if, as is to be expected, profit-maximizing logic leads industry leaders to operate at an output level to the left of the low point on their average cost curves. What in another context is an undesirable situation is here behavior which helps the economy as a whole to avoid an undesirable situation.

Another line of reasoning might suggest that slack can be achieved by a continuous program of building ahead.[2] This argument is based on the assumption that the industry will inevitably be engaged in continuing expansion, and that by constantly anticipating demand increases sufficiently far in the future, the short-run contingency problem will be solved, as a result of the inevitable over-shoot in expansion always under way. Can this line of reasoning be sustained within the present industry market structure, or would changes in structure and entrepreneurial attitudes be a necessary corollary? Might it

2. "Slack" may be realized also because of the existence of standby facilities, temporarily retired but reactivated during peak periods, and by the reliance on certain amounts of imports. These factors will be discussed in subsequent sections of this study.

also be handled by government construction or underwriting of the construction of the slack facilities?

Decisions as to how much variation there is in this discrepancy between average need for steel output and contingency need in the short run would affect the kind of policy recommendation to be made. For example, if we were to hypothesize a growth rate for the economy of four per cent, and establish that a rate of three per cent in steel would be likely to sustain it (including allowance for secular substitution away from steel), we might nevertheless find that on a short-term basis, spurts of growth would bring the operating rate to something over 100 per cent capacity. We must ask what amount of excess capacity would carry the economy through cyclical variations in growth, without a confrontation of the 100 plus per cent capacity operating rate. If this amount is relatively small, we might reach conclusions about whether the cost should be borne by the private (versus the public) sector different from the conclusions we would reach if the gap, and hence the costs, were large.

Moreover, in connection with the likelihood that oligopoly behavior will leave some capacity unused, we must know whether the unused capacity relates closely to our slack figure, whether it goes beyond the needed standby reserves or, on the other hand, whether it may still leave the economy unprotected. Also, if an oligopoly profit or a subsidy is allowed, we must ask whether this incentive results in the industry having too large an income from an equity standpoint.

Finally, having suggested two normative judgments involved in this discussion—i.e., that unused resources are undesirable, and that inhibitors to growth and stability are to be avoided—we must face a further dilemma. Is it reasonable, given the present market structure of the industry and given the free-enterprise, private decision-making format to which our society appears committed, to assume that market forces will lead to the continuing full exploitation of our productive potential in this basic commodity and that, thus committed to its use, we will not be trapped in "normal" growth periods by its inadequate availability? This is, of course, implied in the policy alternatives mentioned above, where we also make suggestions which point to tampering with the existing market structure. For example, both types of social cost could conceivably be affected by governmental intervention in varying degrees. But do we not also have one further type of social cost, the cost attached to tampering with the fundamental incentive and private decision-making milieu within which we are currently operating? Part of the answer undoubtedly rests upon whether the reduction of one of these costs must be achieved at the expense of one or both of those remaining.[3] Another part of the answer must rest upon a deeper examination of the motivational pattern—the leadership and the market structure of the industry—to see whether expected behavior would lead to the kind of alternatives suggested in the foregoing, or whether, in fact, it is more reasonable to assume that policy changes designed to elicit changes in behavior will fall short of providing meaningful solutions to these problems.

In our society the decision to act is essentially in private hands. Government will (as it has many times in the past) intervene if private activity is in some sense inadequate to or at odds with overriding and broader goals of society at large. Until very recent times, government interference with business activity has, for the most part, focused on activity among competitors and between producers and their customers. In both cases this has been true where the activity was said to be unfair, or in some sense not in conformity with standards of morality or ethical behavior. Only in isolated cases and in fairly recent times (except in the case of public utilities) has the government interfered with decision-making in the investment field, that is, in questions of levels of output and adequacy of capacity, and this has been confined primarily to periods of national emergency. However, as government becomes increasingly aware of the fact that levels of full employment and economic stability

3. There is also a cost to be attached to substitution away from steel as a part of a larger economic program. This, too, must be set against the others.

are interlocked with the flow of investment goods and with capacity adequate to guarantee that this flow will be forthcoming, it is likely that government intervention will occur much more frequently and in a wider variety of cases.

One is confronted with the problem that the decision to act, on the part of entrepreneurs, may not conform to what the government sees to be society's needs in the very spheres wherein the entrepreneur is designated by society as the responsible figure. What does government do when it concludes that such men are behaving in a manner which it deems not beneficial to society's long-run needs? How can government take action and at the same time preserve the basic ideals and forms of a free enterprise society, which relies on private individuals operating freely as decision makers?

This volume is primarily for the policy maker in government and in industrial fields where the types of decisions discussed are relevant. A government policy maker may find this study of interest because he must, in an ultimate sense, come to grips with a particular issue—how can the larger society's needs be fulfilled in certain areas? The businessman should take interest in what is discussed here because assertions are made about his behavior and his degree of responsiveness to so-called social needs and the ensuing effects on the overall functioning of the economy; also because, in the last analysis, this discussion leads to assertions about government intervention in the private sector of the economy which the businessman undoubtedly finds distasteful and presumably would be anxious to avoid. In what follows, we suggest that it is often within the businessman's power to avoid, or at least to minimize, government intervention, through enlightened decision-making. There is, therefore, in this material a prescription for the businessman that will not only affect his actual ability to minimize losses in planning for the future, but also will have implications for what he faces in terms of government interference within his domain.

One might assert that it is not legitimate for the economist to consider the impact on the entrepreneur of government concern with socially or "objectively" determined criteria; that the entrepreneur's role as such precludes such criteria as social or objective performance. It must be said, however, that entrepreneurial reactions and attitudes are so central to our economic organism that it is mandatory for the economist to include the entrepreneur's response to government concern with socially or objectively determined investment criteria in a model of the system. There is no other way to shield analytical schemes from perpetual frustration or eclipse by psychological or institutional factors. Therefore, in addition to its relevance for the future of steel and public policy, this study will endeavor to characterize the interplay of economic variables, environmental conditions, moods, and ideas. Can we support assertions that the decision-making apparatus acts as a fulcrum, that it rests at the center of a play of forces impinging on one another? If this is so, and also if it is so that the decision maker is not only under the constraints of the economic environment (demand conditions, labor and resource availability, state of knowledge, the structure of markets, economic goals, etc.), but is responsive to ideological constraints and to a set of personal goals, then decision-making is shaped not only by economic forces but indeed by the whole cultural environment. For the society to form policy intended to affect decision-making behavior, or for the individual to affect such behavior, analysis must recognize the aims of the entrepreneur, the goals of the society and other exogenous factors, as well as the constraints of environment referred to above. This discussion will then ultimately suggest an assessment of the behavior of individual businessmen in relation to general social needs. Are they performing in a way consistent with maximization of the general welfare? Judgment of performance will depend upon investigation of: (1) the behavior-conditioning environment and (2) the particular record of business leadership in the industry in the past.

Before any alarms are sounded, however, it is necessary to decide whether there is, to any significant degree, a discrepancy between private entrepreneurial goals and larger social goals; i.e., between the conclusions individual entrepreneurs would draw (in our case, for example, as to the needs for expanding capacity in the economy) and what government might conceive to be society's needs. Next it is necessary to find out whether, if a discrepancy is observed, the entrepreneur's prescription is the more or the less desirable one to follow. In some respects, there may be inevitable "superior" wisdom originating from the private sector of the economy. It seems clear, therefore, that in the course of this study a behavioral inquiry to estimate the response sequence of private entrepreneurs in investment decision-making is called for.

The behavioral material presented in this volume may well be the more significant part of any contribution that is made here. It would be gratifying, of course, to feel that the findings and policy recommendations made in the latter sections will be of some constructive use to individuals pondering these matters. Nevertheless, the material which characterizes the motivation pattern of decision makers in iron and steel and which discusses the qualitative aspects of this decision-making within the social context should serve as a reservoir of useful empirical material. It may be combined in another study, perhaps with different analytical consequences, and used in other connections in treating decision-making in the capital goods sector as a whole. Heretofore, studies of investment decisions in industry have largely been quantitative studies that have tended to reconcile proclaimed forecasts of investment plans with plans which materialized in actual outlays; they have concentrated on pinning down—with the use of correlation analysis—the time intervals involved between the various acts. These studies have been integrated with others of business forecasting so that anticipation, planning, and execution data have become more reliable as business indicators; they have been useful, but constitute only one approach to the characterization of investment decision-making in capital goods industry. Through discussion of the institutional structure of the firms involved and through some discussion of the personal histories, backgrounds, and psychological conditioning of individuals who are leaders in their respective firms, further contributing factors in investment-anticipation behavior may be identified.

The considerable amount of material that has appeared in recent years attests to growing interest in and awareness of those aspects of economic theory which have come to be known as economic dynamics. Evidence of this new awareness can be seen in the writings of such men as Harrod, Domar, Hicks, and others.[4] The growth of interest in pursuing this line of inquiry in economic theory was inevitable, because earlier attempts to describe economic relationships were inadequate to cope with what is designated today as the dynamic setting of economic activity. This development has obvious significance for students in the field of business cycles, international trade, and public policy. What has not always been apparent nor clearly pointed out is the importance that this recent literature may have for study of the firm itself and, recognizing the oligopolistic structure of major industry, for the study of industrial sectors as well.[5] Certainly the implications underlying investigations on growth, secular stagnation, and economic dynamics, as such, carry with them suggestions of the importance of inquiries into the activities of individual firms and industries. However, with rare exceptions,[6] there have been

4. Among the discussions on this subject the following may be cited: R. F. Harrod, "An Essay in Dynamic Theory," *Economic Journal, 49* (1939), 14–33, and *Towards a Dynamic Economics* (London, Macmillan, 1948), lecture 3; J. R. Hicks, *A Contribution to the Theory of the Trade Cycle* (London, Oxford University Press, 1950); E. D. Domar, "Capital Expansion, Rate of Growth and Employment," *Econometrica, 14* (1946), 137–47, and "Expansion and Employment," *American Economic Review, 37* (1947), 34–55; N. Kaldor, "A Model of the Trade Cycle," *Economic Journal, 50* (1940), 78-92; T. C. Schelling, "Capital Growth and Equilibrium," *American Economic Review, 37* (1947), 864–76; J. S. Duesenberry, "Hicks on the Trade Cycle," *Quarterly Journal of Economics, 64* (1950), 464–76; S. C. Tsiang, "Rehabilitation of Time Dimension of Investment in Macrodynamic Analysis," *Economica N. S. 16* (1949), 204–17.

5. Appropriate here might be a brief discussion of the inadequacy of some treatments of the dynamic process stemming from the essentially *partial* equilibrium framework employed. The central importance of a general framework will be developed in a later section.

6. Cf. the work of W. W. Leontief, *The Structure of American Economy, 1919–1939* (2d ed., revised, New York, Oxford University Press, 1951) and *Studies in the Structure of the American Economy* (New York, Oxford University Press, 1953). In addition, see R. Eisner, "Guaranteed Growth of Income," *Econometrica, 21* (1953), 169–71; E. Glaser, "The Inter-Industry Economics Research Program of the Federal Government," (Conference on Research in Income and Wealth, New York, National Bureau of Economic Research, 1952), p. 14.

few departures from aggregative analysis in such investigations.

This study attempts to demonstrate that some reconciliation may be achieved between analysis of the less generalized level of economic activity (that is, the firm and the industry level) and aggregative analysis of dynamic phenomena. Material will be developed which is specifically relevant to industry-level (or even firm-level) planning and which will, at the same time, be directly concerned with discussions of the aggregative dynamic economic scheme. It is also our intention to illustrate the strategic role that particular industries play, within the so-called dynamic structure, by the very nature of their relation to some of the principal unknowns in the growth process.

Much of recent literature on economic development has been concerned with discovering how an economy can maintain balance between the use of its resources and its ability to finance expansion.[7] Numerous formulations have been presented which assert that a certain volume of investment must accommodate such parameters as the propensity to save, the productivity of capital goods themselves, and so forth, in order to achieve growth and stability. Solutions have tied the achievement of this goal to a specific rate of growth. Thus, if the expanding society is to realize some crucial rate of net capital formation, it must increase levels of capacity in the capital goods sector. However, though models present targets which "must" be met, they often leave unpursued (perhaps understandably) the question of whether this net investment is likely to be forthcoming.

In earlier discussions of these problems, the system analyzed was closed and in a sense complete. It was maintained that if the motivation to invest falters, then income levels will drop until an equilibrium is approached at a new lower level. Or, on the other hand, if incentives to invest are too strong, then inflationary pressures (or resource limitations) will operate

7. For example, *Hearings* and *Staff Report on Employment, Growth, and Price Levels,* before the Joint Economic Committee, Congress of the United States, 86th Congress, 1st session, and The Rockefeller Brothers Fund, *The Challenge to America: Its Economic and Social Aspects* (New York, Doubleday, 1958).

either to inhibit the implied over-investment or to confront the society with other unwelcome problems. This cyclical model tied investment incentive to expectations, and left them there.[8] However, when one attempts to integrate this picture with long-run growth, one faces the need to pursue the sequence underlying the investment-demand function, for without the specified rate of investment—that is, without a level of expectations holding specifically to the requisite needs of the growth model—the system must by definition fail to expand.

The study which follows has implications for current assumptions regarding response mechanisms both implied and explicit in certain growth models. We will examine data covering decisions to invest related to over-all demand, with a view to describing the relationships of these fluctuating series. But, more than this, we will attempt to relate investment decisions, followed by specific firms within a specific industry, to what is often disparagingly termed the "institutional" context. Our intention in this study is to follow the path backward from the investment demand function; i.e., from expectations back to their formation, and then forward to a discussion of policy formulation which perforce must embrace not only the quantitative complexity of the problem but the difficult behavioral considerations as well.

In Chapter 5 we will examine some reaction patterns from the perspective of the individual entrepreneur. Material concerned with the basis upon which the entrepreneur typically proceeds

8. In his review of Keynes' *General Theory*, Schumpeter expresses his unhappiness with the treatment of expected net proceeds, expected profits, etc. as follows: "The emphasis on *expected* as against *actual* values is in line with modern tendencies. But expectations are not linked by Mr. Keynes to the cyclical situations that give rise to them and hence become independent variables and ultimate determinants of economic action. Such analysis can at best yield purely formal results and never go below the surface. An expectation acquires explanatory value only if we are made to understand why people expect what they expect. Otherwise expectation is a mere *deus ex machina* that conceals problems instead of solving them." See "Review of Keynes' General Theory," *Journal of the American Statistical Association, 31* (1936), 791–95, reprinted in *Essays of J. A. Schumpeter*, ed. R. V. Clemence (Cambridge, Mass., Addison-Wesley, 1951), p. 154.

to investment decisions will be presented, and judgments made as to whether this basis is consistent with, or perhaps responsive to, the so-called externally imposed criteria stemming from the theorist's pronouncements. This will serve as a point of departure for subsequent discussion of how the entrepreneur might ultimately revise the criteria upon which he makes decisions which affect the growth structure of the economy, along lines consistent with what is suggested in the general macrodynamic literature as constituting a necessary part of a stable, expanding economy.

Preceding all this, however, we shall try to determine whether there is any recognizable relationship between the so-called growth requirements of an economy and the particular situation of a particular industry. This must be established before we can attempt to make judgments as to whether or how programs will be instituted to guarantee that certain investment magnitudes will be forthcoming. First, evidence must indicate whether it can be said with confidence that certain investment *must* be forthcoming in terms of specific industries before this growth is possible.

This volume, therefore, is divided into three main parts: (1) discussion of the secular and cyclical role of the iron and steel industry in the course of economic change, with an assessment of the demand structure faced by the industry; (2) study of the responsiveness of the entrepreneur to so-called externally imposed demand on the industry; and (3) alternative policies appropriate to solving some of the questions posed. These parts will be treated as follows in subsequent chapters. Chapter 2 recounts some aspects of the historical relationship between iron and steel, investment behavior, and over-all changes in output and income in the national economy. Chapter 3 contains a formulation of the dynamic economic structure, and an attempt is made to characterize the role of the industry in the course of economic change. Chapter 4 is devoted to analysis of demand in the iron and steel industry linked to the requirements of the national economy. It covers former attempts to estimate demand in the industry, and then presents a possible context for demand analysis. Chapter 5 deals with the question of entrepreneurial responsiveness to what might be termed objective demand analysis. The volume concludes (Chapter 6) with a discussion of criteria for policy decisions on a macroeconomic level as well as on the industry or firm level, in the light of the historical and theoretical evidence presented in the body of the study.

TWO THE RELATIONSHIP OF CHANGES IN THE INDUSTRY AND IN THE NATIONAL ECONOMY

THE CAPACITY ADEQUACY CONTROVERSY AND THE ACCOMPANYING ARGUMENTS

One of the most widely discussed proposals in the President's State of the Union message of January 1949 concerned the nation's steel-producing capacity. Mr. Truman recommended that, if necessary, Congress should authorize government loans to expand plant capacity in an amount which would relieve such shortages as were found to exist. Indeed, he went further, suggesting that Congress authorize the actual construction of facilities if private enterprise did not itself undertake needed construction.[1]

Reaction to the President's proposals—as might have been expected—was not unanimous. Leon H. Keyserling, with surprising reserve, said that he believed the members of the Council of Economic Advisers supported Mr. Truman's steel program.[2] Senator Homer Ferguson, on the other hand, was pleased to observe that it was just another "step toward socialism."[3]

1. President Truman suggested that the Congress authorize "an immediate study of the adequacy of existing facilities in short supply, such as steel." He went on to propose that "if found necessary Congress should authorize Government loans for the expansion of production facilities to relieve such shortages, and to authorize the construction of such facilities directly if action by private industry fails to meet the need." *State of the Union Message to the Congress of the United States,* January 5, 1949.

2. "Truman's Advisers Uphold Steel Plan," *The New York Times* (January 19, 1949), p. 30.

3. "Steel," *Time* (January 17, 1949), p. 77. The American Iron and Steel Institute responded with a large advertisement proclaiming the industry's part

Differences of opinion as to the adequacy of steel capacity were not only a post-World War II phenomenon: considerable attention was devoted to this subject in the 1930s. Statements made in 1949, subsequent to the President's message, however, were numerous, and they reflected the general nature of positions characteristically held.

The arguments of government economists, who felt that steel capacity had been inadequate, centered on the principle that a full employment economy was a necessary part of "the promotion of the general welfare." The justification for government concern rested on three main contentions: (1) that bottlenecks in meeting pent-up consumer demand to a large extent might be due to steel shortages, which in turn could exert inflationary pressures; (2) that national security depended on an adequate supply of steel for armaments and for foreign aid commitments; and (3) that a future economy of full employment and full production was dependent upon adequate future steel producing capacity. Steel to meet security needs as well as to meet pent-up demand might be classed as short-run considerations; however, it was asserted that per capita needs for steel had long-range implications for a stable economy.

The logic followed the following lines: it is insufficient to assume that the economy's future needs for steel products will follow the same pattern as in the past, or to expect that past peak levels of demand will keynote future demand patterns. Rather, it is more reasonable to assume that past trends in rates of increase in steel demand will dominate in the future. More-

---

in post-war production: "America's steelmaking capacity is now 96 million tons per year—larger than ever before in war or peace. This great capacity enabled American steel companies to produce 55 per cent of all the world's output of steel in 1948. Capacity is still growing. By the end of next year new facilities now under construction will raise capacity above 98 million tons. Cost of expansion completed since the war's end and now in progress will exceed 2 billion dollars. You have read of record breaking production throughout most lines of manufacturing in 1948. You know that shortages of goods are disappearing. Employment is at a peak. But do you realize that production and employment records could not have been attained without an enormous supply of steel—the basic metal of industry? As always throughout its history, steel is expanding its capacity in step with the growing needs of the country." *The New York Times* (January 10, 1949), p. 13.

over, calculations must be made on a per capita basis so that the allowance includes consideration of population changes as well as consumption pattern changes. From this, it was argued that significant rises in annual demand for steel should be anticipated, because of increases in both population and per capita consumption.

The economists taking this position did make certain assumptions regarding the overall employment conditions they expected would exist in the future. Full employment was assumed. In fact, because of it, "peak demand" was implicitly equated with normal or average demand. The result was to move the concept of capacity utilization into the background. One hundred per cent utilization was implied.

Using this basis of demand estimate and the assumption of full employment, these men had little doubt as to the urgency of building steel plants to meet the demand. As a matter of fact, it was argued that if plant capacity were not raised to the level of adequacy cited, the whole structure would fall apart. Demand would not be met; investment programs necessary to sustain income and employment could not be undertaken; and the result would be that the limits of capacity in the steel industry would be setting the bounds of growth and prosperity in the economy in general.

Recognition of the potential bottleneck position of the steel industry was reflected in the report of the Iron and Steel Committee of the International Labor Organization which met in Stockholm in 1947. Speaking of the importance of a high level of investment from the point of view of employment policy, "and in order to eliminate as rapidly as possible the shortages of capital goods which create bottlenecks and thereby hinder both an increase in employment and the production of consumer's goods," the Committee evaluated estimates of future steel capacity. They concluded that stability programs for the economy must include a reversal of usual concern over the possible effect of a slump in economic activity on the steel industry. "Instead of inquiring about the contribution which the general economy can make to steel, by way of a maintained demand, the question becomes: what contribution must the steel industry make, by way of supply of materials, if the economy as a whole is to be fully employed?"[4]

Supported by this appraisal of the need for adequate steel capacity, some economists turned to examine the expansion programs of the iron and steel industry. Some, particularly before the Korean war, found the industry stubbornly unwilling to meet what seemed to them to be minimum full employment capacity needs. Those claiming that capacity had been inadequate maintained that the observed reluctance of the industry to expand stemmed mainly from undue emphasis on its experience in the 1930s. Then, capacity utilization fell below 20 per cent. The opinion of this group was that it seemed unlikely that the future would resemble the experience of the past. True, if there were expansion, steel might suffer degrees of excess capacity at certain times. Nevertheless, consideration had to be given to prescribed levels of capacity needed throughout the economy, if bottlenecks were to be avoided in an economy experiencing growth.

Industry leaders, of course, disagreed sharply with these views.[5] Industry estimates of demand were based on past peak

4. International Labor Organization, Iron and Steel Committee, *Regularization of Production and Employment at a High Level,* Report No. 2, second session (Stockholm, 1947), pp. 57, 67.

5. Not only with the sequence of ideas, but with some of the intellectual background as well: "With the little band of Keynesian economic planners who have their little 'compartments' in every department and bureau of the Government in Washington, the greatest crime of which the steel industry is guilty, is the crime of failing to keep abreast of the 'managed' expansion of the economy. Thus, we have the fantastic paradox in which the steel industry is being belabored by the Fair Deal economists for not expanding its productive facilities while the Fair Deal politicians, by their constant demonstration of hostility toward the industry, have effectually blocked the only natural course of such expansion under the American free enterprise system. Last week, a spokesman for the Department of the Interior repeated this now familiar charge. 'The industry's planned 10-per-cent expansion is not enough,' he said, adding that 'an increase of 25 to 30 per cent will be needed by the end of 1952.' While such figures as these have been put forward repeatedly by the self-appointed planners of the national economy in the Department of the Interior, the Department of Agriculture and elsewhere, this writer, for one, has never seen one scintilla of evidence that the authors recognized any basic inconsistency in Washington's attitude toward the problem. They have never, in fact, indicated that they were aware that there was any relationship between prices and profits, on the one hand, and the rate of industrial growth on the other." E. H. Collins, "Facts on Steel Production," *The New York Times* (October 23, 1950), p. 34.

years as maxima. Projections presuming economic growth were not used. In part, this reflected concern with providing a reasonable return on investment to stockholders. But indeed most significant to industry was its assumption regarding the economic future. The industry was profoundly impressed by its experience of the 1930s. Moreover, its conservative attitudes had roots in the experience of an even earlier period. Just following World War I, considerable pressure to expand had been brought to bear on the industry. It was asserted then that the industry could look forward to ten years of clear prosperity, for the world was starving for steel. Charges were made that the industry was deliberately creating an artificial shortage of steel. This was early in 1921. Industry leaders recalled how wrong these claims had been. Before the end of 1921, steel production dropped to less than 35 per cent of capacity.

This experience, combined with that of the 1930s, when steel production averaged only 40 per cent of capacity, led the steel entrepreneur to ask what assurances really existed that the whole pattern would not repeat itself again. He noted that profits depend on per cent of capacity in use, and that when there is a situation such as existed in 1937-38, when weekly production varied from a high of 84.5 to a low of 23 per cent, it is not likely that the "fair return" on invested capital would be forthcoming.

To further complicate the picture, in the later 1930s the industry had faced accusations of having too much capacity. The N.R.A. Code expressly forbade the introduction of any new productive capacity in the industry: "It is the consensus of the Industry that, until such time as the demand for its products cannot be adequately met by the fullest possible use of existing capacities for producing pig iron and steel ingots . . . none of the members of the Code shall initiate the construction of any new blast furnace or open hearth or Bessemer steel capacity."[6] In the course of the Temporary National Economic Committee (TNEC) hearings, it was suggested that the redundant capacity of the steel industry had been one of the major causes of

6. U. S. National Recovery Administration, "Code of Fair Competition for the Iron and Steel Industry," Article V, section 2, p. 13.

the depression and the widespread unemployment of that decade. In fact, in a singular example of reversal of roles, the United States Steel Corporation was found pleading the necessity for excess capacity:

> True, the steel industry had a large amount of unused capacity during the recent depression years, but this is reasonable and to be expected in an industry with capacities that are rigid and immobile and whose rate of operations is so controlled by the tremendous cyclical fluctuations in the demand for steel. If the industry is to have the facilities to supply the peak or near-peak demand, it must have idle capacity during the periods of lower demand. . . . Quite conceivably, with any capacity less than it presently possesses, the steel industry would become a bottleneck and prevent full and normal recovery.[7]

It was with this background of paradoxical experiences that the industry faced the pressures of the period immediately preceding World War II, and its leaders expressed their fears of over-expanding. And it was in this situation that both the government and the industry brought forth, in considerable detail, various estimates of capacity and demand, and plans for expansion. Steel executives assured the government that adequate plant existed; yet we will find that shortly thereafter a ten million ingot ton expansion program was launched.

Even as the war effort proceeded, pronouncements emanating from leaders in the steel industry predicted a period of maladjustment and overcapacity as a result of the "over-expansion" to meet war needs. It is worth quoting at some length from two addresses of Walter S. Tower, then president of the American Iron and Steel Institute, during this period; they reflect not only the general frame of reference of leadership in the industry, but also the arguments that were to persist for some time after the close of the war. In 1942 Tower stated:

7. U. S. Congress, *TNEC Hearings,* 76th Congress, 3d Session, part 26, January 23–25, 1940 (Washington, Government Printing Office, 1940), p. 13909.

For your own industry, however, there are certain aspects of the future which can be thought of in concrete terms.

With frozen costs and inflated taxes, a decline in prices or in volume of production when the war ends could quickly bring acute trouble to steelmakers. What adequate provision is being or can be made to provide an effective cushion against inevitable shock? For the aftermath of large-scale war has ever been reaction, unemployment, falling prices, and economic readjustment. Having prior realization of what may happen, is there any way to soften the blow when it falls?

We know pretty well what the capacities and the facilities of the industry are likely to be at the end of this year, at the end of next year, and on to the end of 1944. Those are already well laid down in an orderly program of growth and dictated expansion. In a space of three or four years the industry will have anticipated all the expansion and evolution that normally could have been expected in a generation. Other industries will have stood still or actually lost ground through complete inactivity in their accustomed lines.

Whatever the date of the war's ending, the steel industry will find itself equipped with capacities and facilities for which there will be little, if any, need. There is no reason either in present fact or in past experience to suppose that all those facilities can be used in meeting peacetime requirements.[8]

And in 1944:

Prospective postwar demand for steel has been a subject for much speculation. Disciples of the so-

8. Walter S. Tower, "Steel in a Year of War," an address before the 51st general meeting of the American Iron and Steel Institute (New York, May 21, 1942), p. 11; in *Year Book of the American Iron and Steel Institute, 1942* (New York, American Iron and Steel Institute, 1942).

called 'national income' theory of business activity, using 'free wheeling' methods of calculation, turn up some truly fantastic figures. As much as 120 million tons a year is one result of applying their process. Using another approach, the advocates of the 'full employment' idea of postwar conditions profess to see no decline from the peak of wartime levels. Cold facts as to what major industries have needed in the past, and may reasonably be expected again to need for civilian markets, suggest a figure in the range of 65 to 70 million tons of ingots in good years after the war stops.

That last figure is substantially more than actual consumption in any past peacetime year. It would represent an operating rate of about 70 to 75 per cent of present capacity. Yet at prevailing relations between costs and prices of common grades of steel products, that volume of business would be unprofitable for the average producer. . . .

One hardly needs point out that any such gap as 15 to 20 million tons between usable capacity and expectable market demand will bring its own series of problems. Conspicuous among those problems will be the official attitude toward government-owned facilities for which there is no private buyer. The only sound policy toward such facilities will be to shut them down as soon as shrinking war demand leaves available supply greater than current civilian requirements. Any other policy is certain to involve an element of Federal subsidy, directly or indirectly, to keep such plants going. The question then will be: are we going to have government-owned plants taking business away from privately owned plants?[9]

9. Walter S. Tower, an address before the 53d general meeting of the American Iron and Steel Institute (New York, May 25, 1944), pp. 4–6; in *Year Book of the American Iron and Steel Institute, 1944* (New York, American Iron and Steel Institute, 1944), pp. 29–36.

At the end of the war and by 1949, however, pessimism had been tempered and plans were under way to expand plant further.[10] It is true that much of the net effect was nullified by the necessary scrapping of worn-out and obsolete equipment kept in service during the war to meet the emergency situation. Once again, the stage was set for another round of the capacity controversy. The President's State of the Union message that year and the reactions to it, referred to at the beginning of this chapter,[11] established the tone for the heated debate that was to follow in the months preceding the Korean War.

An interested group was the Steel Subcommittee of the Senate Small Business Committee.[12] They were anxious to determine whether a shortage of steel was responsible for squeezing small fabricators and finishers out of the market because of their inability to meet a "non-competitive" steel price. Here, as before the war, in Senate War Investigating Committee testimony, the opposing sides lined up with reams of statistical material supporting their respective positions. And as before, different conclusions were the logical consequence of the different assumptions regarding the nature of the future economic climate as well as the difference in method used to estimate the nature of future demand.

Again in 1950 hearings were held which concerned the question of adequate steel capacity, this time before the Subcommittee on the Study of Monopoly Power of the House Judiciary Committee. There Ernest T. Weir, chairman of the National Steel Corporation, stated:

10. Planned new capacity then included increases of 3.5 million tons of ingot making capacity, nearly 3 million tons of new coke oven capacity, 3 million tons of new blast furnace capacity, and 5 millions of added finishing capacity.

11. Cf. also "Steel Men Defend Industry's Output: Faster Increase of Capacity Injudicious They Say, in View of Abnormal Demand," *The New York Times* (January 12, 1949). In this article, Ernest T. Weir, chairman of the National Steel Corporation, is quoted as remarking: "It is obvious that the huge immediate need for steel is abnormal and temporary. It would be foolish and damaging to the United States to build permanent capacity in proportion to this passing situation."

12. See citations from this testimony in Chapter 4, p. 107.

> There has never been a steel shortage under normal conditions in my experience. The steel industry had always had capacity exceeding current demand. . . . In the past fifty years, the public requirements for steel has amounted to an average of only 70% of steel capacity.

And at a later portion of his testimony:

> Now, I have pointed out here that we have lost 29 million tons of steel through labor trouble; we lost 10 million beginning with last fall. . . . I say that before the end of the year there will be a surplus of steel available for New England and every other place in the United States. There would be a surplus of steel available for New England today if we had not lost this ten million tons. There is no shortage of capacity for steel.[13]

In answer, Louis H. Bean, economist in the office of the Secretary of Agriculture, stated:

> Steel capacity is now at a record level. For the whole year of 1950, in view of current expansion programs, it will amount to about 100 million ingot tons exceeding the wartime peak of 96 million tons. But to say that steel capacity is now at an all time high does not indicate its adequacy.[14]

Later, in the period of expansion following the outbreak of the Korean War, executives in the steel industry worried over the possibility of over-expansion. The impetus to expand attended the federal government's action to stimulate private investment through accelerated amortization for tax purposes of "all or part of the cost of needed expansion over a relatively

13. U. S. Congress, House of Representatives, *Hearings before the Subcommittee on Study of Monopoly Power of the Committee on the Judiciary*, 81st Congress, 2nd Session, December 19, 1950 (Washington, Government Printing Office), p. 66.

14. Ibid., p. 67.

short period rather than over the normally longer life of the facilities involved." Indications of the impact of this write-off provision on plans to expand plant can be seen from these figures: at the outbreak of World War II capacity was at 81.6 million ingot tons; on December 7, 1942, it was approximately 88.5 million tons; at the close of the war, it was 95.5 million tons; when Korea was invaded expansion had reached a capacity of 100.6 million tons. Another perspective of this expansion is revealed in figures published in the Department of Commerce report on which industries had taken advantage of the amortization program launched in the later period, as compared to the comparable program of the period 1940–45. Whereas iron and steel accounted for only 12.1 per cent of the total value of the certificates of necessity issued during World War II, in the Korean conflict period, through April 13, 1951, iron and steel accounted for 44.3 per cent of the total value of projects approved by the government.[15]

Though they did raise their expansion targets,[16] steel executives continued to express concern over expansion programs.[17] Their contention was, as it had been after World War II, that they would find themselves in a very serious situation once the defense program was modified. They stated that, on the basis of what they considered normal demand, the industry might well be over-expanded. Yet the industry was clearly taking

15. "Accelerated Amortization and Private Facilities Expansion," *Survey of Current Business* (May, 1951), pp. 11–13.

16. In his Economic Message to Congress of January 12, 1951, President Truman exhorted further expansion from the steel industry: "The steel industry 'must' raise the productive capacity above its present level of about 103 million ingot tons a year to support the defense effort and sustain the civilian economy. The industry's own plans reportedly contemplate an expansion to 110 million ingot tons, but an Administration source expressed confidence that the industry would accept the goal of 120 million ingot tons, to be attained in the next three or four years, which the Council of Economic Advisors told the President was necessary." "Truman asks Heavy Tax Rise to Pay for 140 Billion 2-Year Arms Cost: Approves 262,502 Increase in Troops; Pleads for Unity: Gains Urged in Steel and Power Output to Back Defense," *The New York Times* (January 13, 1951), p. 1.

17. C. M. White, president of Republic Steel Corporation, was quoted commenting on the steel industry's expansion program: "On the basis of normal steel demands, the steel industry is now over-expanded. If one takes into consideration the additional capacity either under construction or an-

advantage of accelerated amortization.[18] But might not too much capacity be constructed with too high rates of amortization allowances, and was not there an apparent inconsistency in the industry's estimate of future demand and its action on expansion? Perhaps, acknowledged C. M. White, president of Republic Steel, "but which is better," he asked, "to have the Government build it or have the companies build it through accelerated depreciation? That's the real question, isn't it? Where will it serve the country best?"[19]

Domar, referring to entrepreneurial reaction to what are seen as prospects for sustained levels of demand, comments on the effect of expectations on plant expansion thus:

> Past depressions do exert a profound influence on business thinking, and an assurance that they will not recur would undoubtedly brighten up the future and make many marginal products worth undertaking. If, in addition, businessmen could confidently expect a growing economy, this effect would

---

nounced, this is shown by the 1949 record, when steel production dropped almost 11 million tons as a result of cancellations and the decline in the production of consumer goods. Another indication is the fact that on the basis of 117.5 million ton annual capacity (scheduled to be in operation by the end of next year), the per capita capacity will be 1,482 pounds. This is an increase of 111 pounds over present per capita and is more than is needed to take care of normal demand." "Steel Men Worry on Over-Expansion: Leaders Say Capacity May be Pushed to Excessive Heights by Defense Hysteria," *The New York Times* (February 4, 1951), p. F.1.

18. And of opportunities to obtain loans from the government as well. "One prominent steel official disclosed recently that there are scores of applications for government funds to finance proposed new steel facilities. Last week the newly formed Gilbraltar Steel Corporation said it plans to build a new integrated steel mill of 800,000 tons annual capacity near Detroit, provided it can obtain a loan of $90 million from the Reconstruction Finance Corporation. The proposed New England mill to be located near New London, Conn. is another facility which may get Federal financial aid for 1 million tons of ingot capacity, if it fails to interest an operating steel company in building the mill. The Sheffield Steel Corporation also disclosed plans for a $71 million expansion program, contingent on Government financing. Earlier the Lone Star Steel Company obtained Government loans totaling $73,425,201 to permit it to undertake a large expansion in Texas." Ibid.

19. "The 'If' in Steel Expansion," *U. S. News and World Report* (December 1, 1950), p. 38.

> be much stronger. The recent publication by the American Iron and Steel Institute, *Background Memoranda—Steel Capacity,* is an excellent demonstration of this point. The purpose of this memorandum is to show that the country possesses sufficient steel capacity, both relative to the peak (peacetime) year of 1929, and to the peak demands of the most important users of steel taken individually. And the only growth admitted into these prognostications was the growth of population. If a sufficient number of our industries make their plans along these lines, we will end up with some fifteen or more million unemployed. But so long as the probability of future depressions is great, can we expect different plans to be made? And the sad joke about all this is that after the depression does come, these planners will justly congratulate themselves on their remarkable foresight![20]

The controversy had not been resolved as a result of the change in plans subsequent to the war in Korea.[21] The issues were reargued at intervals throughout the '50s as swings of economic activity brought first peak demand, then periods of relatively low capacity utilization. The controversy persisted in spite of the fundamentally broader planning horizon of the industry (to be discussed in Chapter 5) and despite the change to an essentially peace-time planning base.[22] It is true that arguments were modified by more expansive planning attitudes on the part of management, but the fundamental points that had been at issue remained unchanged. Moreover, as we entered the '60s, there was confusion and disagreement as to the means of

20. E. D. Domar, "The Problem of Capital Accumulation," *American Economic Review, 38* (1948), 786.

21. Steel-making capacity at the beginning of 1953 was 117,547,470 net tons annually, of which 17 million tons were installed during the Korean War. See "U. S. Mills Set New Steel Output Peak," *The New York Times* (December 13, 1953).

22. See, for example, F. Fogarty, "Needed Soon: More Materials Capacity," *Architectural Forum, 109* (1958), 140–41.

establishing verification of the assumptions underlying points at issue. In order to get some perspective for later discussion, we must turn to data covering changes in the relevant variables. At the outset we must try to shed some light on the essential economic character of the steel industry viewed as an element in the structure of the national economy.

QUESTIONS REGARDING THE BASIC INDUSTRY CONCEPT

In a post-World War II article on the British iron and steel industry, D. L. Burn, an expert in the field, questioned whether the distortions that existed in the industry itself were of such fundamental significance to the economy that correction of them was of paramount importance.[23] He pointed out that "rationalization" of the industry would perhaps achieve a closer approach to optimum operating conditions, but that accompanying sociological and political repercussions would negate, in their turn, the possible advantages that could be gained by revisions of the structure of the industry itself. He concluded his remarks by questioning whether the industry was so "basic" after all, and added that perhaps the pressures of other considerations should dominate the scene.

Though this particular example has its setting in Great Britain, it effectively serves to raise the question of the fundamental relationship of the iron and steel industry to an economy as a whole. The word "basic" in Burn's statement was not intended, of course, to indicate an industry close to a natural resource source of productive activity; i.e., he was not using the term to imply likeness to such industries as water power, mining, or the like. Rather, he was questioning whether the general level of economic activity and the existence of higher standards of living and full employment are in any way essentially dependent on adequate capacity or optimal efficiency in the iron and steel industry in a given economy or at the disposal of a given

23. D. L. Burn, "Recent Trends in the History of the Steel Industry," *The Economic History Review, 17* (1947), 95–102.

economy. In this sense, "basic" implies a dependency relationship.

The difficulty encountered in the use of the term lies in the ease with which a transition is made from viewing the apparent closeness of movement between series to the assertion that concomitance is invariable.[24] One expression of this ready general-

24. In connection with a plea for the maintenance of protection to iron and steel manufacturers, we find an early reference to the potential bottleneck factor of inadequate steel production facilities in the economy in the following, taken from "On the Product and Manufacture of Iron and Steel," The General Convention of the Friends of Domestic Industry, assembled at New York, October 26th, 1831. Reports of Committees, signed by Order of the Committee by B. B. Howell, Secretary. Asserting that the protection afforded the industry resulted in relative price stability in the economy, and in fact a reduction in prices over time, arising from domestic competition, the Committee report goes on to say:

> This comparative stability, so important to the success of all well regulated industry, was due, exclusively, to the domestic supply, which effectually protected the consumer from the foreign speculator, who could otherwise have controlled this market, and produced here the same disastrous consequences that ensued in his own.
>
> If such has been the result of protection upon the general market of the country, its effect would have been still more striking, when examined with reference to particular, but most important districts. Our western brethren, the hearty pioneers of our country, were restrained and limited, in their contest with the wilderness, by the difficulty of obtaining, on almost any terms, these articles so indispensable to their success, in every state of their arduous enterprise. The second statement of prices, included at the end of the Report, exhibits the prices of iron of various description at different periods, in Pittsburgh and Cincinnati, the great marts of the west. Comment can scarcely be necessary upon the facts there disclosed. The decline in price (in some instances more than one-half) has been in exact proportion with the stability given to the domestic manufacture, by additional imposts on the foreign, until it has reached a point that now enables the mechanics of the first mentioned city, that Birmingham of America, to enter into successful competition with those of almost any other quarter, in the fabrication of nearly every article of necessity, and in one, justly esteemed the proudest effort of human ingenuity, they have attained a degree of perfection which enables them to challenge comparison with the skills and experience of any nation whatever (pp. 3–4).

Here, in addition to conventional infant-industry and self-sufficiency arguments in support of protection, we find an early nineteenth-century analog to the arguments put forth in the post-World War II period, where the claim was made that the development of the economy would be impeded by inadequate production facilities in this very same industry.

ization is found in the statement of Oscar Chapman, former Secretary of the Interior, in testimony before the House Subcommittee on the Study of Monopoly Power.

> The steel industry is essential to the American economy as the steel reinforcing parts are to a large dam and power house. Every characteristic American business depends upon ample supplies of low-cost steel. . . . So close is the relationship of steel and our economy that they even seem to have the same coefficient of expansion. Historically, the production of steel has matched our climb to pre-eminence among industrial nations of the world. In 1909, when our Nation's output of goods and services was valued at about $80,000,000,000, at present prices, we produced about 27,000,000 tons of steel. This required 51,000,000 tons of iron ore. In 1929, peak of economic activity during the 1920's, we mined 73,000,000 tons of ore and produced 63,000,000 tons of steel. This contributed to a total national output of goods and services of about $160,000,000,000. In 1948 we produced more goods and services than ever before in peacetime, valued at $260,000,000,000. Corresponding steel output was almost 90,000,000 tons, and that of iron ore about 101,000,000 tons.
>
> Note that changes in steel output and changes in gross national product, measured by a constant price level, tend to be closely matched. From 1909 to 1948, for example, steel output slightly more than trebled and so did gross national product, at constant prices. During the past several generations, a growth in national product of about $3,000, at 1949 prices, has matched each ton of increased steel output.[25]

In looking at data concerning the relationship between

25. U.S. Congress, House of Representatives, *Hearings before the Subcommittee on Study of Monopoly Power of the Committee on the Judiciary,* 81st Congress, 2d Session, Serial No. 14, part 4A (Washington, Government Printing Office, 1950), pp. 69–70.

movements in the value of U.S. Gross National Product and U.S. steel production for the period 1919–50, we can see an apparent degree of parallelism between the two series.[26] In Table 1 and Figure 1, agreement appears in all years except four. High points coincide in the years 1929, 1937, 1944, and 1948. Lows coincide in the years 1921, 1938, and 1949. There is some indication that steel lagged G.N.P. in 1923–24 and also in 1926–27. In both cases, there was little change in G.N.P., and in the latter year of each pair steel output fell. In the large dip of the early '30s, the steel drop apparently led the fall in G.N.P., with G.N.P. hitting bottom in 1933 rather than 1932. However, other figures for G.N.P. can be found where the dip does occur in 1932. The discrepancies in both 1919–20 and 1946–47 can be attributed to "outside" factors, particularly labor trouble in the industry and related industries. In Figure 1, growth of G.N.P. can be seen to be at a rate of 3.2 per cent per year, and, for the same period, growth in steel output is 3 per cent per year.[27] Therefore, it appears that both in terms of turning points and secular trend, the series are quite close. It is, of course, impossible to substantiate lines of causation, but it may be noted that a relationship does exist between the output of the steel industry and changes in G.N.P., which closely coincides with variations in the marginal capital coefficient over the cycle.[28] The fact that *steel* in ratio to *changes in output* bears a resemblance to *changes in capital stock* in

26. A section devoted to an historical account of the origin and growth of the iron and steel industry in the United States seemed an unnecessary repetition of material handled fully elsewhere, and therefore has not been included here. For more complete historical accounts, see V. S. Clark, *History of Manufactures in the United States* (New York, McGraw-Hill, 1929); C. M. Daugherty, M. G. de Chazeau, S. S. Stratton, *The Economics of the Iron and Steel Industry* (New York, McGraw-Hill, 1937); E. C. MacCallum, *The Iron and Steel Industry in the United States* (London, King, 1931); E. B. Alderfer and H. E. Michl, *Economics of American Industry*, 3d ed. (New York, McGraw-Hill, 1957).

27. Both a straight line trend and an exponential trend were fitted to each of the series. In the case of G.N.P., the exponential trend was clearly the better fit. Though for the 32 years graphed, a straight line fit appeared to be slightly better for the steel series, knowledge of what preceded suggested that the exponential was really the better.

28. See Table 6, p. 59.

TABLE 1. *Data Concerning the Relationship between Movements in the Value of U.S. Gross National Product and U.S. Steel Production, 1919–50*

Year	*G.N.P. (billions of dollars) 1939 dollars*[a]	*Steel ingots (millions of net tons)*[c]	*Deviations from straight line trend, G.N.P.*[d]	*Deviations from straight line trend, steel*[e]	*Exponential trend relatives, G.N.P.*[f]	*Exponential trend relatives, steel*[g]
1919	66.5[b]	38.8	18.8	8.8	119.3	119.4
1920	62.7[b]	47.2	11.9	15.5	109.0	141.0
1921	56.4[b]	22.2	2.5	−11.2	95.1	64.4
1922	61.7[b]	39.9	4.7	4.8	100.8	112.3
1923	70.7[b]	50.3	10.6	13.5	111.9	137.5
1924	70.8[b]	42.5	7.6	4.0	108.6	112.7
1925	74.0[b]	50.8	7.7	10.6	110.0	130.8
1926	77.2[b]	54.1	7.8	12.2	111.2	135.2
1927	77.4[b]	50.3	4.9	6.7	108.1	122.1
1928	80.8[b]	57.7	5.2	12.4	109.3	135.9
1929	85.9	63.2	7.2	16.2	112.6	144.5
1930	78.1	45.6	− 3.7	− 3.1	99.3	101.2
1931	72.3	29.1	−12.6	−21.3	89.1	62.7
1932	61.9	15.3	−26.1	−36.8	73.9	32.0
1933	61.5	26.0	−29.6	−27.8	71.1	52.8
1934	67.9	29.2	−26.3	−26.3	76.1	57.6
1935	73.9	38.2	−23.4	−19.0	80.3	73.1
1936	83.9	53.5	−16.3	− 5.4	88.3	99.4
1937	87.9	56.6	−15.6	− 4.0	89.7	102.1
1938	84.0	31.8	−22.6	−30.5	83.1	55.7

TABLE 1—*Continued*

Year	*G.N.P. (billions of dollars) 1939 dollars*[a]	*Steel ingots (millions of net tons)*[c]	*Deviations from straight line trend, G.N.P.*[d]	*Deviations from straight line trend, steel*[e]	*Exponential trend relatives, G.N.P.*[f]	*Exponential trend relatives, steel*[g]
1939	91.3	52.8	−18.4	−11.2	87.5	89.7
1940	100.0	67.0	−12.8	1.3	92.9	110.5
1941	115.5	82.8	− .4	15.4	103.9	132.6
1942	129.7	86.0	10.7	16.9	113.1	133.7
1943	145.7	88.8	23.6	18.0	123.2	134.0
1944	156.9	89.6	31.7	17.1	128.5	131.2
1945	153.4	79.7	25.1	5.5	121.8	113.3
1946	138.4	66.6	7.0	− 9.3	106.5	91.9
1947	138.6	84.8	4.1	7.2	103.3	113.6
1948	143.1	88.6	5.5	9.3	103.4	115.2
1949	142.3	77.9	1.6	− 3.1	99.7	98.1
1950	153.0	96.7	9.2	14.0	103.8	118.5

[a] 1929–50 data from *Survey of Current Business,* January and February 1951.

[b] Obtained by using link relatives based on G.N.P. series published in J. A. Krug, *National Resources and Foreign Aid,* Department of Interior (Washington, Government Printing Office, October 9, 1947), which were in 1947 dollars. These were chained onto Department of Commerce series for 1929–50, yielding 1919–28 figures for G.N.P. in 1939 dollars.

[c] American Iron and Steel Institute, *Annual Statistical Report* (New York, 1949), p. 30. 1950 figure, *Survey of Current Business,* February 1951.

[d] Regression equation: $Y = 97.3 + 3.1X$; origin: January 1935, X units = 1 year.

[e] Regression equation: $Y = 57.2 + 1.7X$; origin: January 1935, X units = 1 year.

[f] Regression equation: $Y = 92.05X^{1.032}$; origin: January 1935, X units = 1 year.

[g] Regression equation: $Y = 52.26X^{1.030}$; origin: January 1935, X units = 1 year.

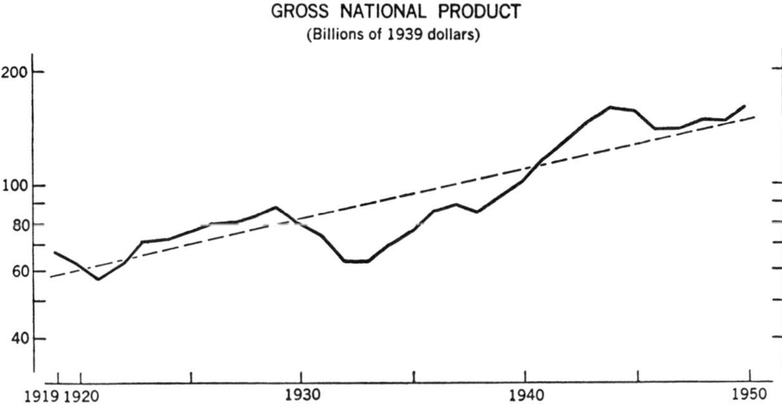

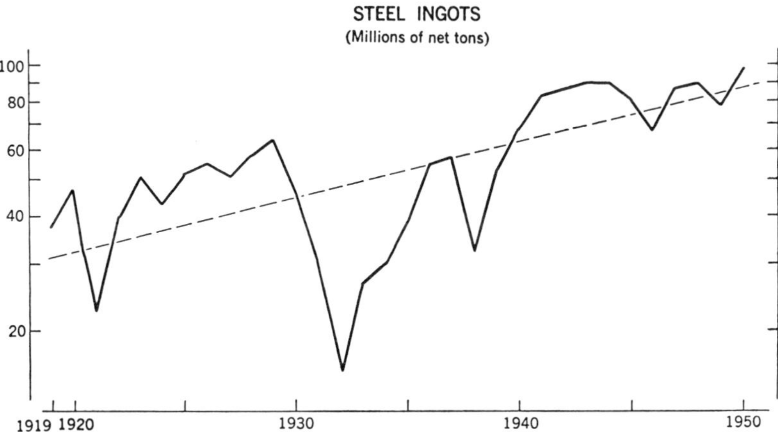

Figure 1. U.S. Gross National Product and U.S. Steel Production, 1919–50

ratio to *changes in output* suggests there are grounds for identifying steel production with investment activity in the economy. Such relatedness does little, however, to aid in the evaluation of the so-called basic position of the industry.

This assumption regarding the industry, nevertheless, has led students of business cycles for some time past to consider iron and steel production figures as a reliable reflection of general levels of business activity. Before moving to a more categorical assessment of the role of the industry, let us look at the use of industry data as "indicators" of activity in the economy.

PREVIOUS STUDIES OF INDUSTRY SERIES

Industry production figures were used to forecast business conditions as early as 1884, when Samuel Benner of Cincinnati published his "Benner's Prophecies." His work involved the use of a lagged relationship and therefore can be referred to as a forecasting device rather than a mere reflection of change. He used iron production figures, claiming: "The changes of the ups and downs in prices and cycles in the iron trade are periodical and not haphazard, and succeed each other in a gradual and natural order."[29]

A later, more systematic use of industry conditions as an economic indicator was formulated by General Leonard P. Ayres of Cleveland, Ohio. In 1923 he set up a "Blast Furnace Index," and was credited as the first exponent of the doctrine that "as iron goes, so goes the market." Ayres' reasoning was based on the assumption that because it cost so much to set a furnace in operation, one would only be started up if there were orders on hand for several months production. In 1925 he stated his theory as follows:

> Over a long period of years it has been true that the average percentage of blast furnace operation, taking good years with bad, has been about 60%.

29. S. Benner, *Benner's Prophecies of Future Ups and Downs in Prices* (1884), cited in *Steelways* (American Iron and Steel Institute, January, 1950).

> This 60% level has continued to represent about average conditions despite the changing trade practices of the iron and steel industry, and despite the increasing size and capacity of the furnaces. Now a careful study of these percentages over past years, and a comparison with the records of changing prices in the stock market, indicates that when the line showing the percentage of active blast furnaces has been above the average, and turned down so as to cross below that level, the prices of stocks are usually at about their low point for that market and soon begin to rise. Moreover, when the blast furnace percentage line has been below the sixty level for some time, and has risen so far in a period of business recovery that it crosses the sixty level on the way up, stock prices are usually near their high point and shortly begin to decline.[30]

Whatever the relative validity of the rationale applied to the empirical findings, Ayres' approach does illustrate reliance on iron and steel industry data for insights into general market conditions. His experience also serves to point out how vulnerable such analyses are to changes in conditions; for in spite of his claim that changes in the industry would not affect his index, it was not long before a method for banking fires in blast furnaces was developed. This made it possible for an active furnace to "mark time" between orders without shutting down. Another change that affected the index was the considerable abandonment of furnaces during 1925 and 1926. The total number of furnaces was reduced from 403 on January 1, 1925, to 369 on July 1, 1926.[31] The fact that the abandoned furnaces were old and less efficient meant that the output of 60 per cent of the remaining furnaces was higher than 60 per cent of the former number of furnaces. Ayres noted: "Month by month it

30. L. P. Ayres, *Business Bulletin* (Cleveland, Cleveland Trust Company, December 15, 1925).

31. H. B. Vanderblue and W. L. Crum, *The Iron Industry in Prosperity and Depression* (Chicago, A. W. Shaw, 1927), p. 23.

looks to us as if the old 60 per cent level which has been a good guide for many years threatens to be less useful in the future. I think that it is pretty clear that for the last two years, at least, the 60 per cent level is too high to constitute a satisfactory normal period."[32]

A more systematic analysis of the iron and steel industry in relation to general economic activity was undertaken by Warren M. Persons in 1921.[33] He noted that the Census of Manufactures of the United States classified manufacturing industries into fourteen groups. Of these groups "iron and steel and products leads all others, whether the groups be ranked according to capital investment, amount of wages paid, or value added to material by the manufacturing process."[34] Persons compared the annual physical volume of production indices for ten groups of manufactures with the consumption of pig iron for the period 1899 to 1919. He found that the range in iron consumption moved from 67 per cent of the norm in 1908 to 119 per cent of the norm in 1906 and 1916, and observed that the minimum and maximum points for all groups occurred in the same years. He claimed that pig iron consumption was a very sensitive index.[35] He noted that a monthly study for 1919–21 showed larger swings for iron because production had been affected by strikes in the fall of 1919. He also pointed out that the decline in consumption of pig iron late in 1920 was forecast by recession elsewhere. All in all, Persons' reaction to the iron series as an indicator was favorable.

A different emphasis is found in the study of iron and steel production and its relationship to the economy made by Vanderblue and Crum in the 1920s.[36] They focused on a representative figure for industry performance. After establishing the role played by iron and steel in contemporary business activity, especially with regard to the railroad, construction,

32. *Business Bulletin.*

33. W. M. Persons, "The Iron and Steel Industry During Business Cycles," *Review of Economic Statistics, 3* (1921), 378–83.

34. Ibid., p. 378.

35. Ibid., p. 381.

36. Vanderblue and Crum, *The Iron Industry* (see note 31).

and automobile industries, the authors turned to a statistical analysis of trends and cyclical change. In their treatment of cyclical movements of pig iron production in the United States they eliminated both secular and seasonal variations from the data, used monthly figures, and expressed them as trend relatives. They assumed the long-run trend to be linear. They did the same thing for cyclical indices of steel ingot production and came to the following conclusion:

> The influence of changing business conditions upon the output of the two principal branches of the iron and steel industry is surprisingly similar, both as regards the timing of the changes of direction of the curves of cyclical indexes and the extent of the swing from trough to peak. The two curves, despite occasional differences in level and direction of movement, move very close together. Because of the similarity, it is clear that the statistics of pig iron production are, on the whole, a very satisfactory rough index of conditions in the steelmaking branch of the industry as well.[37]

Here, as with Ayres, the reasoning falters with changes in technological conditions. This is strikingly true in iron and steel because of the marked shift in the role of pig iron in the steel production process. Pig iron demand is greatly affected by the use of scrap in open hearth and electric furnaces.[38] Unlike ingot production in the Bessemer converter, where pig is the primary input material, open hearth and electric steelmaking employ the use of pig iron in ratio to certain amounts of scrap metal. The differences in the more important inputs can be seen in Figure 2. This ratio can be varied as conditions of relative availability of the two materials vary; and it is this

37. Ibid., p. 22.

38. The Bessemer process involves air blown with great force up through a white-hot molten mass of pig iron in a converter; the open hearth is a reverberatory furnace where hot gases are passed over the surface of molten metal composed of a certain percentage of scrap and a certain percentage of pig iron; the electric furnace melts the charge in the furnace by electric arcs.

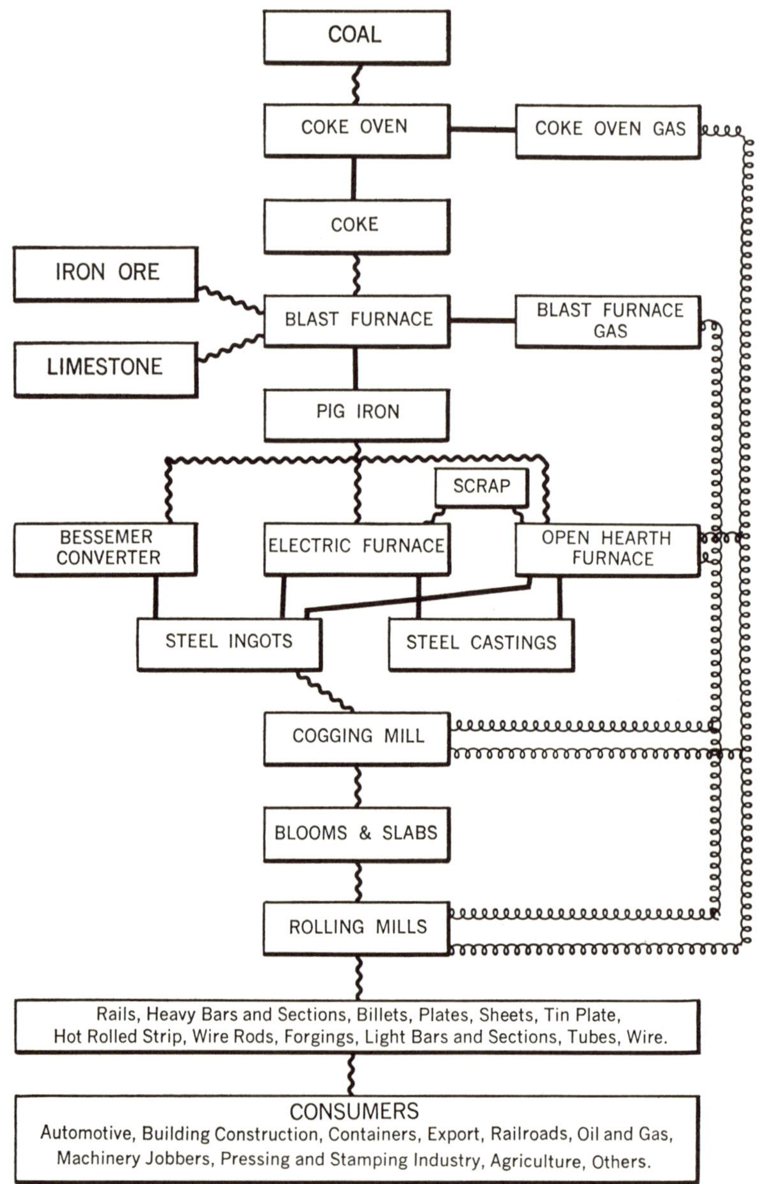

Figure 2. Flow Chart of Iron and Steel Production

fact which introduces a non-linearity in the relationship of pig iron production to steel production. Of course, for this to be of significance, a definite majority of steel producers must shift from a process where pig iron almost completely dominates the picture to one where a flexible pig to scrap ratio becomes possible. This was just the case. In the Bessemer process, the pig to scrap ratio is approximately 9 to 1; Bessemer converter production accounted for approximately 66 per cent of the total of ingots and castings in 1900. It fell to 55 per cent in 1905, 36 per cent in 1910, 26 per cent in 1915, 21 per cent in 1920, 15 per cent in 1925, with the balance over these years being filled for the most part by open hearth steel.[39] In 1940 industry capacity was 88.3 per cent open hearth, 8.2 per cent Bessemer, and 3.5 per cent electric. Combining this information with the fact that the pig to scrap ratio in electric furnaces is 5 per cent to 95 per cent and in open hearth furnaces is usually 50–50 (but can vary with scrap almost completely dominating the input picture)[40] indicates the general weakness of Vanderblue and Crum's assertion. I emphasize this point because it illustrates so clearly how changing technology can render certain economic assumptions useless. The focus on pig iron production as substitutable for or representative of industry production as a whole does hold, perhaps, for the period studied, 1902–23. Even for this period, however, it is not truly indicative because the significant shift from Bessemer to open hearth took place before World War I.

Changing demand patterns for the output of the industry as well as changes in the rate of growth of different sectors of the industry also contribute to the discontinuities which make "well-established" indices suspect. Vanderblue and Crum did illustrate this condition (Table 2). Yet, at the same time, they used a sort of rule of thumb as the basis for representing pig iron production. They took the percentage of the total number of furnaces in blast at any one time (similar to Ayres) and set

39. Vanderblue and Crum, *The Iron Industry*, p. 11.

40. G. Dunn, "Report to the President of the United States on the Adequacy of the Steel Industry for National Defense" (Washington, Office of Production Management, February 22, 1941), p. 15 (processed).

this up as a reflection of the general condition of the industry. It is difficult to understand the justification for such generalization in the face of the apparent factor shifts within the industry at the time.

TABLE 2. *Trends of Iron and Steel Production, 1902–23*[a]

Type	*Central value*[b] *(normal monthly production for January 1, 1913)*	*Annual increment*[b]	*Annual rate of growth*[c]
Pig iron	2,249.3	66.1	2.94
Steel ingots	2,297.8	104.0	4.52

[a]Vanderblue and Crum, p. 22.
[b]Unit: 1,000 gross tons.
[c]Unit: 1 per cent.

Vanderblue and Crum devote some attention to the role of scrap. It is discussed, however, in terms of scrap prices as an index of conditions in the iron and steel industry. Here again, because of lack of recognition of the possibility of using scrap and pig in different ratios in the open hearth furnace, the authors attribute meaning to scrap price movements which may not necessarily hold true. For example, during World War II when scrap was unavailable, we saw the ready substitution of larger amounts of pig iron for scrap in open hearth furnaces. Though it is true that as demand for steel increases and production increases one might expect scrap prices to rise, the substitution possibilities indicate that perhaps the industry's demand for scrap is more price elastic than the authors realized at the time their book was written. This reasoning applies especially in situations where steel producers have some control over their source of pig iron and do not have such close control over purchased scrap. In contrast to this, Vanderblue and Crum assumed that scrap consumption could be easily accelerated when the price of basic iron rises considerably above that of scrap. Any use of a variable such as scrap price movements must be carefully qualified in terms of "availability in good times" as opposed to the more likely greater availability in "bad times." In addition, such questions as the life expectancy of durable goods and changes in the scrap export-import picture must be considered.

Vanderblue and Crum also touch on the general character of demand for iron and steel in the early '20s. A basis for comparing the shifts in demand is provided, for example, in the 1924 and 1939 figures (Table 3). Here again we find grounds

TABLE 3. *Consumption of Steel: Percentages of Estimated Total Steel Output*

Consumer	*1924*[a]	*1939*[b]
Railroads	28	9.3
Building and construction	18	13.1
Automotive industry	10	18.1
Oil, gas, water, and mining	9	5.5
Export trade	6	6.5
Food containers	4	9.4
Machinery	3.5	3.8
Agriculture	3	1.9
All other	18.5	32.4[c]

[a]Vanderblue and Crum, p. 147.
[b]Standard and Poor's *Industry Surveys,* June 27, 1947, section 2, p. S3–6.
[c]Included are Pressing and Stamping Industry (3.6 per cent) and Jobbers (15.6 per cent), not specified in the 1924 listing.

for being wary of reliance on some single industry variable as an enduringly satisfactory indicator. This early study, and others that followed, were untroubled by the interdependency of industrial markets. They provided an essentially static, compartmentalized analysis. As if to emphasize the point, we find that in their conclusion Vanderblue and Crum cast considerable doubt on the usefulness in forecasting of any information on steel consumers or suppliers. For them the problem was reduced to the interpretation of certain series arising from within the industry itself. When limited to such series, they suggest that the industry does reflect general conditions of prosperity and depression.

Several other studies which have included reference to changes of iron and steel conditions over the cycle should also be mentioned briefly. One is Kuznets' *Secular Movements in Production and Prices*. He includes figures on the rate of percentage increase and average deviation from secular trend of a number of U.S. industrial series. Steel is covered from 1865 to 1924. The figure for the average value of total deviations places the industry second among twelve industries covered, and cyclical fluctuations rank the industry first among twelve industries.[41] J. M. Clark has included attention to iron and steel series in his business cycles study.[42] He finds that the reference cycle and the specific cycle are very close and therefore concludes that pig iron and ingots agree "most closely" with the general business cycle. He examines pig iron over thirteen cycles and ingots over three cycles and notes a more rapid increase in the early stages of expansion. In observing the movement of stocks, Clark notes that movement depends on "nearness to consumer"; for example, pig iron falls as business rises, but sheets rise as business rises. Burns and Mitchell also include a pig iron series in their study for the period 1879 to 1933.[43]

To return for a moment to the earlier point that the industry's data have long been used for forecasting, we should note that its reputation for this purpose is still in good standing. In 1950, for example, an article appeared in a publication of the American Iron and Steel Institute entitled "Do Your Own Forecasting." In it, the author made the point that steel production "is still going strong as a guidepost for profits." He justified its use in this capacity on the following grounds:

41. S. Kuznets, *Secular Movements in Production and Prices* (Boston, Houghton Mifflin, 1930), p. 269. For a series, 1865–1924, on Crude Steel Production, including ordinates of line of trend and relative deviations from ordinates of trend, see pp. 366–67.

42. J. M. Clark, *Strategic Factors in Business Cycles* (New York, National Bureau of Economic Research, 1935), especially pp. 26, 50, 230.

43. A. F. Burns and W. C. Mitchell, *Measuring Business Cycles* (New York, National Bureau of Economic Research, 1947), p. 518.

> About 40% of all manufacturers use steel as a raw material. Furthermore, steel is produced only on special order. Its production represents an estimate by manufacturers as to what and how much the public will buy in the next few months. This explains why steel has been called a 'bell-wether' industry. Monthly steel production figures are important if you want a factual answer to that very common question: 'How's business?'[44]

With these surveys and statements in mind, let us examine some specific data.

PROBLEMS ASSOCIATED WITH TREATMENT OF DATA IN VIEW OF TREND-CYCLE INTERACTION

The following material will be presented in roughly the same sequence that it was analyzed, and some of the difficulties and obstacles that accompanied its analysis will be recounted. At the outset, a twofold task was set and certain restrictions were placed on the data to be employed. First of all, it was considered necessary to establish, within the framework of the data to be used, a clear picture of the relative movements of Gross National Product and the iron and steel output series. It appeared desirable to substantiate empirically (via time series) what has been assumed theoretically by previous observers to be the relationship between a capital goods material, such as steel, and fluctuations in the national economy. A close relationship was expected, a priori. Nevertheless, it was uncertain whether a marked difference in this relationship would appear in a comparison of output measures (such as G.N.P. and steel production) in contrast to a comparison of levels of investment or income and steel demand (such as net capital formation and domestic consumption of steel).

44. *Steelways* (New York, American Iron and Steel Institute, January, 1950), pp. 6–9.

Therefore, the first objective was to establish the closeness, over time, between movements in the steel series and movements in product-income series from a national standpoint. The second was to evaluate the use of these series in extrapolation, attempting to estimate G.N.P. from known steel production figures. The limitations set, as far as the statistical material was concerned, were first, restriction to series that are widely used and readily available, and second, use of series that have a definitional relationship to a priori formulations. It was decided to use series representing the aggregate production picture in one case, and to use more selective series representing domestic steel consumption and investment in another. In an attempt to contrast the comparison between the most "all-inclusive" series with that of the comparison between more "refined" formulations, it was decided in the first instance to compare constant dollar G.N.P. with production of steel ingots and castings, and in the second to compare net domestic capital formation and domestic steel consumption. Elaboration of both of these sets appears below, but before investigating them, it is appropriate to indicate the basis on which series were chosen from those available.

The time period chosen for the G.N.P.-steel production comparison is 1919 to 1950. To have gone back of 1914 would have involved the use of figures which, in the case of steel at least, would represent radically discontinuous changes. Proceeding from the post-World War I date, 1919, seemed to yield a series which characterizes conditions in the more "modern" history of the industry. The series comparison ends in 1950 because the major part of our capacity adequacy discussion focuses on the pre-Korean War period. When using national income data drawn from Kuznets, however, there is a limitation to the period; his series cover the period 1919–38.[45]

As was pointed out in reference to the series in Table 1, movements in G.N.P. and steel production, 1919 through 1950, reflect the degree of closeness that exists between the two series. It was also noted that discrepancies seen both in 1919–

45. S. Kuznets, *National Income and Its Composition, 1919-1938* (New York, National Bureau of Economic Research, 1941).

20 and 1946–47 might be attributed to external factors, particularly labor trouble, both in the industry itself and in related industries. Strikes, of course, are not the only external factor affecting production; strikes in contributory industries or in transportation, shortages of raw materials, especially scrap and ore, manpower shortages due to overfull employment, and so forth—all can disrupt production schedules. However, we must be careful not to dilute the analysis by eliminating years or periods specifically on the grounds that something unusual or exogenous occurred. To do this would leave the series incomplete; one or several of the most significant contributing factors may be removed from the picture. We must then set up some criteria to aid in judging where to classify these "special" events. For example, the duration of their period might indicate whether they were part of the over-all cyclical pattern, or whether their nature was such as to suggest so considerable a change that a new trend, or something similar, might be assumed. Moreover, if there is a valid basic tie between the industry and the economy in general, the occurrence of a strike in the industry should be reflected in transmittal to the G.N.P. series. Timing of events, however, may be so critical that the absence of effect on G.N.P. would not necessarily disprove the "basicness" theory.

In line with the question of trend, certain statements may be made as to its choice and fit. For example, Vanderblue and Crum discuss the requirements of fitting a secular trend line. They point out that one of the requirements of a good trend is that if the trend is plotted on the same chart as the data, "it should trace through the successive cyclical fluctuations in such a manner that at nearly all points of the record the trend line may freely be assumed at the level about which the cyclical fluctuations are, for the time being, taking place."[46] They state, as a second requirement for a satisfactory trend, that it should be capable of projection beyond the available data. They indicate that for the period immediately preceding the beginning of a data series and for the immediate future following its ter-

46. Vanderblue and Crum, *The Iron Industry*, p. 160.

mination, the trend should be a good representation of what would be normal. In discussing the use of curved trend lines, they suggest that convex or concave lines should be used only where a priori reasoning indicates that this type of development is possible. Claiming deficiency of clear evidence in favor of one of the more complex trends, the authors turn to the two simplest—linear and exponential curves. In their view, the straight line offers an apparently good fit; thus they conclude that it is best suited to represent the steel series.

In contrast to this, when I examined the data and considered the "a priori reasoning regarding the statistical series" and which development was most likely, the exponential trend proved to be the more desirable fit for our steel production data. This was true in spite of the fact that for the 32-year period of the actual series, a straight line appeared to be a slightly better fit and yielded, in terms of trend relatives, a higher correlation coefficient, when related to G.N.P. trend relatives.[47] Figure 3 shows G.N.P. exponential trend relatives plotted with both steel production linear and exponential trend relatives.

The choice of the exponential trend was made for two reasons: (1) from what is known of the role that capital goods industries have played in economic development in the United States over the past 75 years—in part a manifestation of capital intensification in production—it seems logical to expect that capital expenditures have increased exponentially, when it is found that G.N.P. has increased exponentially; (2) it can be observed that the exponential trend line and the linear trend line cross each other at two points in the course of the series and, as can be seen in Figure 4, this occurs approximately at the 1920–21 figure and also at the 1946 figure. To make a final judgment as to which is the better fit, it is necessary to see which line, projected both backward and forward, coincides most closely with what actually occurred or what might be

47. There was little question as to the desirability of an exponential fit for the G.N.P. series. Steel production and G.N.P. series with exponential trends appear in Figure 1.

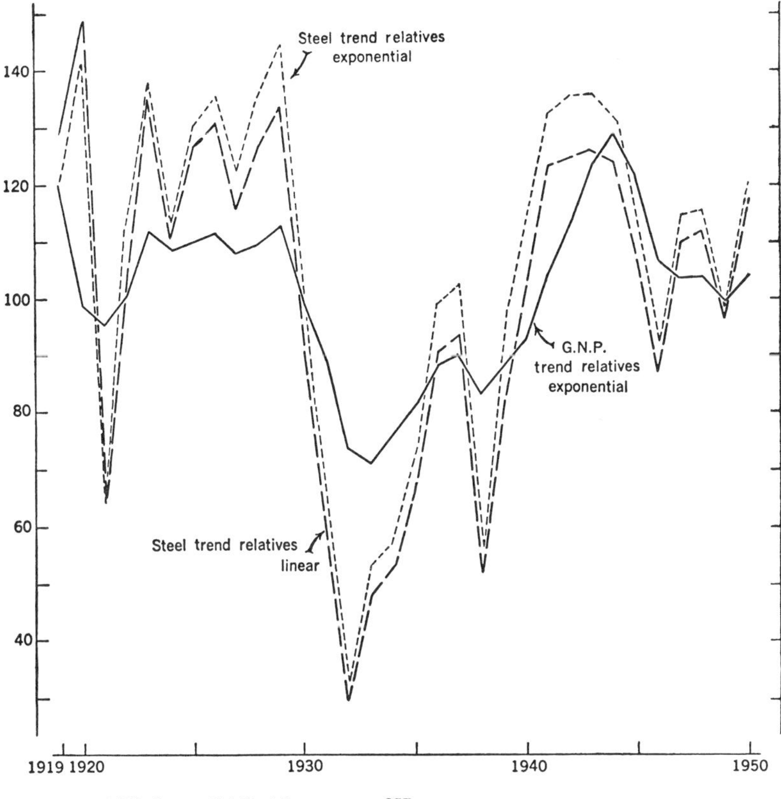

Figure 3. Trend Relatives, 1919–50

expected, as the case may be. Projecting the linear trend back to 1903, for example, gives an estimate of steel production of 4 million tons, and we know that the actual figure was 16.2 million tons (approximate figures); again in 1910 the linear trend suggests 15.3 million tons, and the actual figure was 29.2. In contrast, the exponential trend suggests a production figure of 18.5 (the actual was at 16.2) for 1903. The 1910 figure is 24.9 for the exponential trend (29.2 for the actual). On the "future" side also, the exponential curve is closer to what was realized. It suggests for 1953 something between 100 and 125 million tons, whereas the linear trend suggests an output of

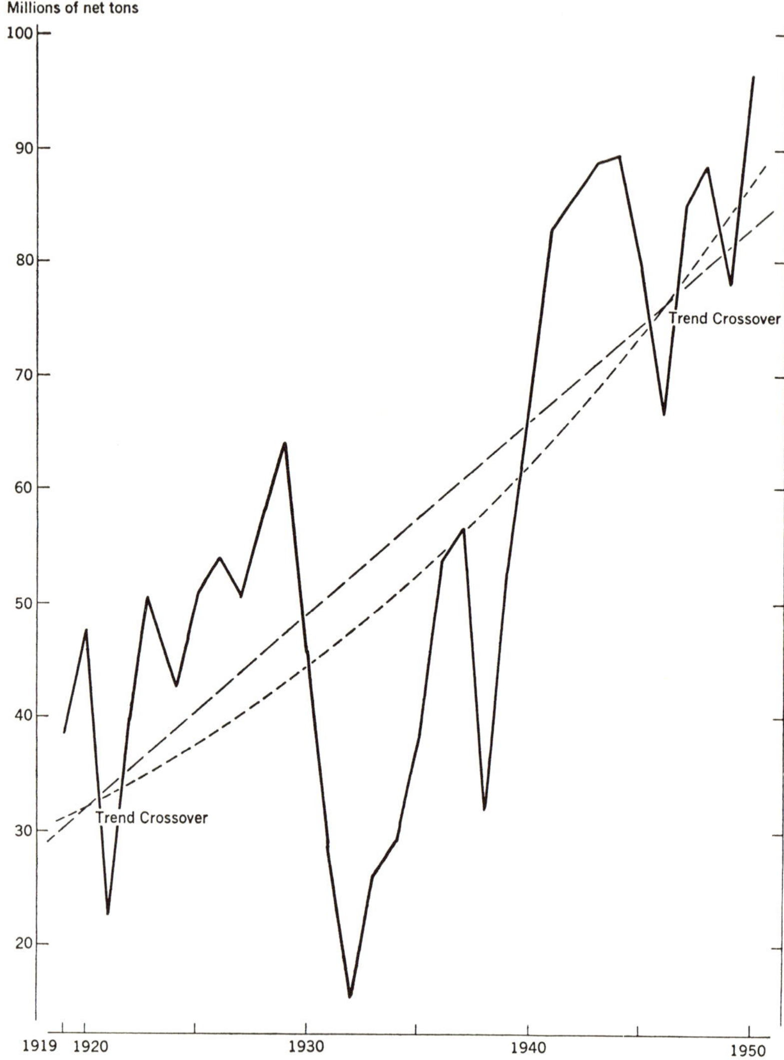

Figure 4. U.S. Steel Production Trend Comparison (ingots and castings in million net tons)

approximately 87 million tons (actual was 111). From this it seems reasonable to conclude that the exponential trend fitted to steel production will satisfy our purposes. It was chosen, therefore, for use in extrapolation.

Before leaving the trend question, there are two additional points to be made. One concerns the implications of trend extraction vis-à-vis the interplay of paths of equilibrium and cyclical fluctuation. The "elimination" of trend is not undertaken without recognizing that interaction between secular and cyclical forces may well be a primary conditioning factor. Fellner has stressed this: "Secular trend lines do not show the equilibrium path along which the economy would be moving in the absence of cyclical disturbances. A reduction in cyclical fluctuations would greatly affect the secular trend."[48] Granting this, however, does not preclude the examination of cyclical fluctuations of the various series; what should be watched is the elimination of trend and then a subsequent ignoring of this step when interpreting final results.

The second point relates once more to the use, as well as choice, of trend for projection.[49] An example of trend choice (using a Gompertz curve) that misfired was that made by the National Association of Manufacturers in their 1941 discussion of government plans for expanding the steel industry. The picture they forecast was an ever-increasing gap, except for inordinate fluctuations, between capacity and demand so long as capacity continued to rise. As a result, they feared overexpansion of capacity. They stated:

> It is observed that even in the peak year 1929, there was a wide gap between production for domestic use and capacity. Since that time a leveling off process in the trend of production and ever increasing capacity have tended to widen the gap. The rate by which production was increasing from year to year was definitely diminishing even before the interruption of the '30s. . . . It might well be that in postarmament years the total demand will not exceed

48. W. Fellner, *Monetary Policies and Full Employment* (Berkeley, University of California Press, 1947), p. ix.

49. An extreme example of the liberties that can be taken with trend projection can be seen in A. A. Mol's "Does the Nation Need More Steel Capacity? Projection of Past Trends Suggests Sharp Production Decline in Years Ahead," in *Barron's Weekly* (January 17, 1949), p. 11.

> approximately 60,000,000 tons of ingots per year, including exports.[50]

The discrepancy between the estimates derived from the use of the N.A.M. projection and the realities of the production picture can be seen by referring to Table 1 (p. 33). Even in 1946, when the economy experienced a severe dip in steel output, the figure exceeded the Association estimate: it exceeded the trend figure by approximately 13 million tons, or about 25 per cent. When more recent production figures are taken, output exceeds trend figures by more than 100 per cent. The Association was wrong in its initial conception of post-war demand patterns; this, of course, dominated their trend selection, and the resulting false picture followed quite logically.

SERIES RELATIONSHIPS

We are now at a point where we may examine a specific exercise in the use of these series for forecasting, as well as for estimating past levels of the one from past levels of the other. The possibilities of refining and "sensitizing" the data will be mentioned below, but first we will examine results obtained from using the G.N.P.-steel production relationship.

An attempt was made to project backward to see if G.N.P. could be estimated from known past steel production data. In this case, data associated with Table 1 were used. The two trend relative series were plotted against each other as indicated in Figure 5 (the regression equations derived appear there also). The method used to obtain estimates of G.N.P. for the years 1909 through 1918 was as follows: the trend lines for both the G.N.P. series and the steel series were extended back to 1909; knowing the actual steel figures, the steel trend relatives were calculated; these trend relatives were substituted in the regression equation, yielding G.N.P. trend relatives, which were then multiplied by the trend values for G.N.P. for the

50. National Association of Manufacturers, *The Iron and Steel Industry in the Defense Economy* (New York, 1941), p. 67.

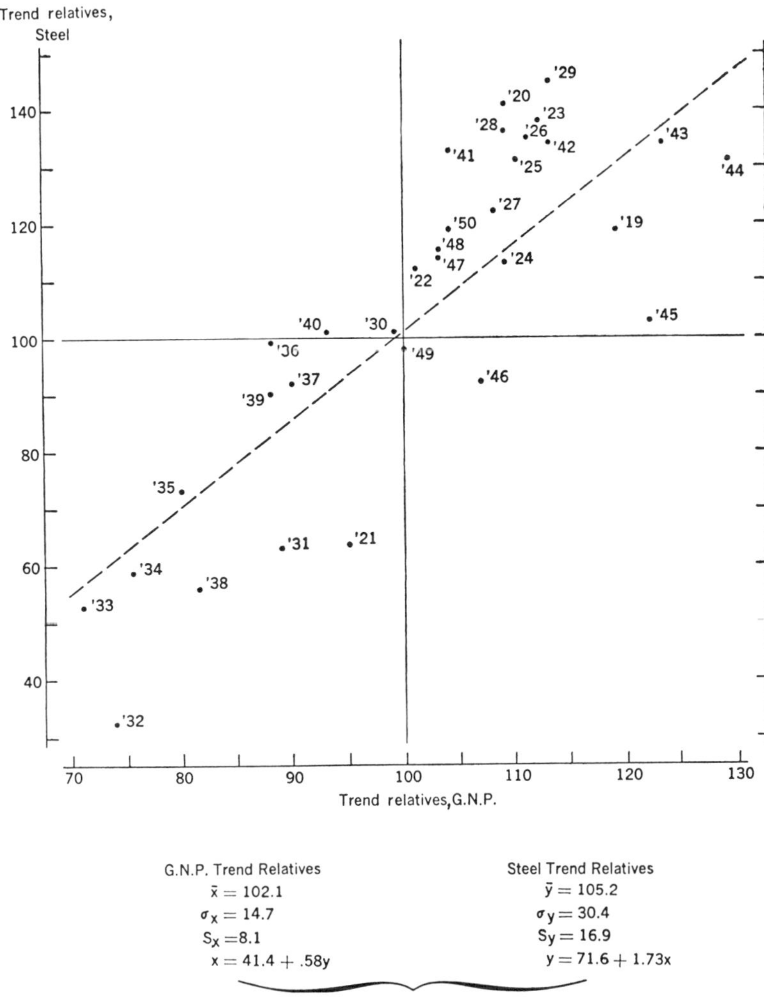

Figure 5. Trend Relatives, Steel versus G.N.P., 1919–50

respective years, thus yielding estimated values for G.N.P. The last step was to find a series of actual G.N.P. to use as a control. This was done by a link relative method, similar to that described in Table 1, note b, page 34. The pertinent data appear in Table 4. A considerable similarity can be noted here be-

Table 4. *Estimated and Actual Gross National Product, 1909–18*[a]

Year	*Steel trend*	*Steel trend relatives*	*Derived G.N.P. trend relatives*	*G.N.P. trend*	*Estimated G.N.P.*	*Actual G.N.P.*
1909	24.1	111.2	105.9	40.7	43.0	45.2
1910	24.9	117.2	109.4	42.0	46.0	48.1
1911	25.6	103.5	101.5	43.4	44.1	49.4
1912	26.4	132.5	118.3	44.7	52.8	52.3
1913	27.2	129.0	116.3	46.2	53.8	54.8
1914	28.0	98.0	98.3	47.6	46.8	53.0
1915	28.9	124.5	113.6	49.2	55.9	52.3
1916	29.7	161.4	135.1	50.7	68.5	60.0
1917	30.6	165.0	137.2	52.3	71.7	61.9
1918	31.5	158.0	133.0	54.0	71.9	63.9

[a]Basis for deriving data contained in text.

tween the estimated value and actual value. This is particularly true of turning points, with two exceptions, 1911 and 1915. In the case of 1915, the upturn in the estimated series preceded the actual upturn by one year. This might be explained by saying that the estimated series is probably "dominated" by activity in the steel industry. The considerable rise in steel production, to a point above its own trend, could be carried over into the estimate of G.N.P. Another factor to consider is possible error from a source suggested by Orcutt, i.e. that this situation may arise because the regression equation was calculated over a period dominated by a cyclical movement of a different character or duration than that which might have been operative from 1909–18. Yet even with this "closeness," we are left with little confidence in the results. Perhaps it is a question of the coarseness of the data.

In an attempt to establish a more sensitive relationship, Kuznets' national income data for 1919–38 and his capital formation series were used, the latter modified to give "net domestic capital formation." These figures were compared with changes in steel *consumption* over the same period. The data appear in Table 5.[51] It can be determined from this that deviations from trend of steel consumption and deviations from trend of national income correlate with a coefficient of .838 (standard error of estimate of 5.11 million net tons), hardly better than what was obtained by checking the closeness of relationship of steel production deviations to G.N.P. deviations. On the other hand, deviations of steel consumption from trend correlates with deviations of net domestic capital formation from trend at a definitely higher level. The r value is equal to .921 (standard error of estimate of 1.61 million net tons).

The steel series seem to be more closely related to investment, as such, rather than to more inclusive aggregative series. This relationship is consistent with the logic of those who assert that capital goods industries are particularly sensitive to fluctuations in general economic activity. This closeness of steel to

51. In this case, linear trend lines were fitted to the national income, net domestic capital formation, and steel consumption series.

TABLE 5. *Data Concerning the Relationship between Movements in the Value of U.S. National Income, Net Domestic Capital Formation, and U.S. Steel Consumption, 1919–38*

Year	*National income (billions of 1929 dollars)*[a]	*Net domestic capital formation (billions of 1929 dollars)*[b]	*Steel consumption (millions of net tons)*[c]	*Deviations from trend, national income*[d]	*Deviations from trend, N.D.C.F.*[e]	*Deviations from trend, steel consumption*[f]
1919	57.0	6.0	33.2	− 7.3	− .6	− 6.7
1920	58.4	6.0	40.7	− 6.5	− .3	.8
1921	56.5	1.6	19.2	− 9.0	−4.4	−20.7
1922	60.8	3.6	37.3	− 5.3	−2.1	− 2.5
1923	70.7	8.0	47.2	4.0	2.6	7.4
1924	71.7	4.9	39.9	4.4	− .2	.2
1925	73.9	8.6	48.5	6.0	3.8	8.8
1926	79.0	8.6	51.4	10.5	4.1	11.7
1927	79.6	7.5	47.6	10.5	3.3	8.0
1928	81.1	6.7	54.5	11.4	2.8	14.9
1929	87.1	9.6	59.4	16.8	6.0	19.9
1930	79.9	3.7	43.1	9.0	.4	3.6
1931	69.3	− .1	28.0	− 2.2	−3.1	−11.5
1932	55.6	−5.6	15.1	−16.5	−8.3	−24.3
1933	56.7	−5.2	25.4	−16.0	−7.6	−14.0
1934	62.1	−4.0	28.3	−11.2	−6.1	−11.0
1935	65.6	.6	37.4	− 8.3	−1.2	− 1.9

TABLE 5—*Continued*

Year	*National income (billions of 1929 dollars)*[a]	*Net domestic capital formation (billions of 1929 dollars)*[b]	*Steel consumption (millions of net tons)*[c]	*Deviations from trend, national income*[d]	*Deviations from trend, N.D.C.F.*[e]	*Deviations from trend, steel consumption*[f]
1936	75.0	6.0	52.6	.5	4.5	13.3
1937	80.8	6.5	53.0	5.7	5.3	13.8
1938	79.0	2.2	29.5	3.3	1.3	– 9.7

[a] Simon Kuznets, *National Income and Its Composition, 1919–1938*, p. 269.

[b] Ibid. Obtained by subtracting "Change in Claims Against Foreign Countries" from "Capital Formation." Percentage share figures on components of capital formation as part of national income are given in Kuznets, p. 272.

[c] Obtained by subtracting net steel exports from steel production. Net steel exports are exports minus imports; both are minus respective scrap movements. Steel production does not include castings. Data from *Annual Statistical Report* of the American Iron and Steel Institute.

[d] Regression equation: $Y = 70.3 + 0.6X$; origin: January 1929, X units = 1 year.

[e] Regression equation: $Y = 3.6 - 0.3X$; origin: January 1929, X units = 1 year.

[f] Regression equation: $Y = 39.5 - .04X$; origin: January 1929, X units = 1 year.

the investment aggregate leads one to examine the possible closeness of growth in the industry in particular to growth in the economy in general, as implications here suggest interpretations of movement of the series broader than the merely cyclical sense. Figure 1 shows the closeness of the growth rates of the economy and the industry respectively over the period 1919–50. As has been pointed out, the G.N.P. trend rises at a rate of 3.2 per cent and that of steel production at 3.0 per cent. It seems reasonable to assert that though quantitative amplitude comparisons seem to break down, comparisons of turning points and growth rates prove to be rather more satisfactory. Statements as to the line of causality would be ill-advised, but it would not be unexpected to find that a relationship existed between steel output and *changes* in G.N.P. This corroborates parallels observed between the industry's output and the investment sector of the economy.

The data in Table 6 yield a comparison between changes in capital stock in the economy and consumption of steel over the same period. It is true that introduction of lag considerations could be expected to alter the results, but the year to year concurrent series do present a picture which is likely to be improved, rather than worsened, in an analysis which includes whatever interval may exist between the use of steel and the manifestation of increases in output in the economy. Nevertheless, it is apparent from the last three columns that some marked similarity does exist. The $\Delta K/C_s$ relationship remains stable with the exception of the period 1931–1935. During this period it appears that activity was increasing in the steel industry at a rate faster than that of investment in the economy generally. But except for this abnormal situation, the range of the ratio does not exceed the limits of .100 and .185; furthermore, this period does cover other years of marked fluctuation. The fact that a resemblance exists between steel in ratio to changes in output, and between changes in capital stock in ratio to changes in output, suggests that there are valid grounds for identifying steel production with investment activity in the economy, but much more must be known before the necessary qualifications to this statement can be made.[52]

TABLE 6. *Capital Stock Change–Steel Consumption Relationship, 1920–38*

Year	ΔO[a]	*Steel consumption* (C_s)	$\frac{C_s}{\Delta O}$	$\frac{\Delta K^b}{\Delta O}$	$\frac{\Delta K}{C_s}$
1920	1.4	40.7	29.1	5.4	.185
1921	− 1.9	19.2	−10.1	− 1.5	.148
1922	4.3	37.3	8.7	.9	.103
1923	9.9	47.2	4.8	.8	.167
1924	1.0	39.9	39.9	5.5	.138
1925	2.2	48.5	22.0	4.0	.182
1926	5.1	51.4	10.1	1.7	.168
1927	.6	47.6	79.4	13.3	.168
1928	1.5	54.5	36.4	4.9	.134
1929	6.0	59.4	9.9	1.7	.172
1930	− 7.2	43.1	− 6.0	− .6	.100
1931	−10.6	28.0	− 2.6	− .02	.004
1932	−13.7	15.1	− 1.1	(−,−) .4	.364
1933	1.1	25.4	23.0	− 4.5	−.196
1934	5.4	28.3	5.2	− .7	−.134
1935	3.5	37.4	10.7	.1	.009
1936	9.4	52.6	5.6	.6	.107
1937	5.8	53.0	9.1	1.1	.121
1938	− 1.8	29.5	−16.4	− 1.8	.109

[a]Net national income differences, cf. Table 5, pp. 56–57, column 2.
[b]Computed by E. Kuh and W. F. Mennick, Business Cycles Seminar, Harvard University, 1951. ΔK = net change in capital stock.

In summary, then, there appears to be considerable doubt as to the usefulness of steel series as indicators or forecasters of levels of business activity. This is particularly true in regard to amplitude, and it is quite unlikely that effective use can be made of steel series for forecasting purposes, as the timing of the peaks and troughs of both the steel series and the general series

52. A relaxation of the assumption that industry output flows exclusively to investment activity is discussed on pp. 67–68.

coincide. Not much in the way of strikingly reliable results were obtained by extrapolating backwards to calculate levels of G.N.P. from known levels of steel output. The estimates derived from the regression formula were satisfactory until a year in which steel pulled far out of line. Then the domination of the steel situation seemed to throw the G.N.P. estimates off. Therefore, here again, skepticism as to its effectiveness arises. Whatever is true of the difficulty of projecting into the past is true of the future; but with the latter, the accompanying pitfalls are even more serious. Future estimates based on trend projections are insulated from the impact of the vagaries of economic and social life. Furthermore, even if it might be said with assurance that the two series were tied together in some significant sense, the problems associated with swings in the investment series, as such, would remain. Of course, there are situations in which planners and policy-makers are pressed for decisions, when even the crudeness of trend projection can be of great service. The use of steel series does not, however, seem to hold much promise for true refinement of prediction procedures. At best we might say that whatever effectiveness as a predictive indicator steel series have had in the past, it seems likely that they will be less meaningful in the future. Complex lag structures plus the complications of "administered" inventory accumulations suggest why. However, it is not in this area that one must confine his search or rest his case concerning the "basicness" of the industry.

The positive result that does seem to have come from this survey is the appearance of steel industry data in some significant relationship to investment data for the economy as a whole. This is, in a sense, substantiation of what has been said regarding the so-called basic position of the industry, its sensitivity to the complex pattern of the demand it faces, and its natural responsiveness to fluctuations in business activity. The object of the following chapter will be to establish a more generalized characterization of the role of the industry in the context of economic change.

THREE THE ROLE OF THE IRON AND STEEL INDUSTRY IN THE COURSE OF ECONOMIC CHANGE: A THEORETICAL FORMULATION

A brief historical account of the variations in output of the iron and steel industry has led us to a point where we should look at some of the theoretical relationships linking the factors under discussion. The following examination of industry parameters, in the context of specific formulations of economic growth and cyclical change, will attempt to provide an analytical basis for evaluating the role of iron and steel in the course of economic change.

To begin with, a brief statement is needed to delimit the meaning of "change." What is "dynamic" in economics, comparative statics, and so forth, of course, cannot be treated in what follows.[1] It is important, however, to place conceptual limits on our use of the terms "economic change" or "dynamic economic activity." Leontief has said:

1. Cf. S. Kuznets, "Static and Dynamic Economics," *American Economic Review, 20* (1930), 426–41; J. M. Clark, *Preface to Social Economics* (New York, Farrar & Rinehart, 1936), chap. 6; F. H. Knight, *The Ethics of Competition* (New York, Augustus M. Kelley, 1951), chap. 6; J. A. Schumpeter, *The Theory of Economic Development* (Cambridge, Mass., Harvard University Press, 1949), chap. 1; W. W. Leontief, "Postulates: Keynes' *General Theory* and the Classicists," in *The New Economics,* ed. S. E. Harris (New York, Alfred A. Knopf, 1948), chap. 19, and *Studies in the Structure of the American Economy,* part 1; J. R. Hicks, *Value and Capital* (Oxford, Clarendon Press, 1946), chap. 9, and *A Contribution to the Theory of the Trade Cycle,* chap. 1; R. F. Harrod, "An Essay in Dynamic Theory," *Economic Journal, 49* (1939), 14–33 and *Towards A Dynamic Economics,* lecture 1, and *Economic Essays* (New York, Harcourt Brace, 1952), chap. 13; W. J.

> Within the framework of an explicitly formulated theoretical system, economic change can be explained either as structural change or as a dynamic process. In the first case, the variation of the dependent variables is simply related to the underlying changes in some of the basic data; in the second, the law of change itself is considered as given, i.e. built into the structure of the explanatory scheme. The law of change might, of course, be changing over time; this is the case of structural variation in a dynamic system.[2]

Discussion in this study will for the most part relate to the *dynamic process* as such. Our context will not be confined merely to structural change or variation. On the other hand, there will be no presentation of so elaborate a mechanism as structural variation in the dynamic system.

INDUSTRY OUTPUT, CAPITAL COEFFICIENTS, AND THE DYNAMIC CONTEXT

Classifying certain industries as capital goods industries and others as essentially consumer goods industries may be difficult and in some cases meaningless. In addition to the purely semantic problem, and that of establishing methods of "recognition," there is also the question of the purpose of the whole enterprise. Justification for the establishment of these categories becomes more apparent when discussion centers on the acceleration principle.[3] To point up the relevance of this concept, a

---

Baumol, *Economic Dynamics* (New York, Macmillan, 1951); P. A. Samuelson, "Dynamics, Statics, and the Stationary State," *Review of Economic Statistics, 25* (1943), 58–68, and "Dynamic Process Analysis," in *A Survey of Contemporary Economics*, ed. H. S. Ellis (Philadelphia, Blakiston, 1948), chap. 10.

2. Leontief, *Studies in the Structure of the American Economy*, p. 17.

3. Cf. S. S. Alexander, "Issues of Business Cycle Theory Raised by Mr. Hicks," *American Economic Review, 41* (1951), 861–78; S. C. Tsiang, "Accelerator, Theory of the Firm, and the Business Cycle," *Quarterly Journal*

rather extreme example might be set up. Assume that all of the output of the steel industry goes into the production of capital goods; that is, steel output is channeled exclusively into investment, with replacement forming a consistent part of gross investment.[4] If this is the case, then, there exists a fundamental relationship between the expansion rate of the steel industry and the change in the rate of output of the economy in general. Does the rate of output of the steel industry (O_s) rather than changes in its rate of output ($\Delta O_s/\Delta t$) have the significant linkage to ΔGNP? If this is the case, a given level of O_s and therefore a certain aggregate of plant capital in the steel industry (K_s) is sufficient to sustain ΔGNP, so long as this is constant. Therefore, if the stability requirements of the growth analysis dictated a constant advance in GNP of e.g. ten billion

---

of Economics, 65 (1951), 325–41; J. S. Duesenberry, "Hicks on the Trade Cycle," *Quarterly Journal of Economics, 64* (1950), 464–76; R. S. Eckaus, "The Acceleration Principle Reconsidered," *Quarterly Journal of Economics, 67* (1953), 209–30; R. Fels, "The Theory of Business Cycles," *Quarterly Journal of Economics, 66* (1952), 25–42; B. G. Hickman, "Capacity, Capacity Utilization, and the Acceleration Principle," in *Problems of Capital Formation,* a Report of the Conference on Income and Wealth, National Bureau of Economic Research (Princeton, Princeton University Press, 1957), and "Comment" by F. Modigliani, pp. 450–57. Bibliography of additional material on the controversy concerning the acceleration principle may be found in the articles cited above.

4. Two particular problems associated with the separation of replacement from investment on new capital should be acknowledged here. The first would be delays in or decisions against replacement; if such were the case, casting replacement in a passive role, as is done above, would result in questionable conclusions. Even this might be permissable, however, on the ground that such delays or decisions average out over time—the latter being especially justifiable when interest is focused on steel output for long-run expansion of the economy. The second problem arises if there is a definite shift to capital-saving equipment in an industry heretofore employing much steel-consuming machinery. This could be handled by a shift in weighting or, if in a continuous direction, by correction from a trend line drawn to account for it. Another problem arises from increases in the productivity of replacement equipment. This may affect the analysis if the new machines require the same amount of steel, but at the same time are ordered in lesser number because the demand faced by the consumer industry is unchanging. Increasing the productivity of capital equipment vis-à-vis an expanding market for the consumer-industry output may conceivably result in no change in replacement needs for steel. Here, of course, it is assumed that no changes take place in the amount of steel embodied in the now more productive equipment.

dollars worth of goods and services annually, and the total output of the steel industry went into investment, then a constant O_s would satisfy the situation and K_s would remain constant (i.e. zero ΔK_s). Note, however, that when there is a change in ΔGNP, ΔO_s must be considered, and subsequently ΔK_s. If, as is contended by Domar for example, it is mandatory that GNP increase at a constant rate, then ΔK_s enters the picture in a relatively permanent manner.

This example breaks down on many points, several of which are of particular importance to this discussion. First, there is the question of secular changes in the marginal capital-output ratio. Fellner has indicated that both marginal and average capital-output relationships remain significantly stable over the period 1874–1929 using decade series, and this seems to hold for both middle- and end-of-decade dating.[5] He adds, moreover, that sector data do not bear out the apparent stability of coefficients as well as aggregate data do. He finds, for example, that in the period 1879–88 to 1889–98, the overall K/O ratio rises, as do the sector ratios. In the period 1889–98 to 1919–28, however, whereas the overall K/O shows stability, individual sector ratios show net decreases. Changes in output composition appear to be responsible for this discrepancy. The situation seems to have been one in which manufacturing, with a relatively low K/O, rises, whereas agriculture and mining remain unchanged. Utilities, railroads, and so forth, which have a high ratio, fall, and this fall outweighs other rises. To reconcile these sector changes with the apparent relative stability of the aggregate K/O ratios, it is suggested that the composition of output traced a pattern wherein heavier industries outweighed lighter industries at the beginning as compared with the end of the period surveyed.[6] It is, however, the specific relevance of

5. Cf. W. J. Fellner, "Long-Term Projections of Private Capital Formation," Conference on Research in Income and Wealth (New York, National Bureau of Economic Research, 1951), p. 43 ff.

6. It might be noted here that no value judgment should be implied as to the relative desirability of trends from heavier to lighter industries or trends in the opposite direction. The essentially dynamic element is one which manifests itself in changes in the output structure of the economy in general, and

steel industry output to this secular path that is of interest to us. Sector shifts in the capital-output ratios must not be ignored; the relevance of sector capital coefficients and the advantage of being able to identify them is forcefully demonstrated in the fact that these series differ from aggregate ratios.

The next point concerning the relationship of capital expansion in steel to a sustained growth rate in the economy centers on the fact that the accelerator relationship does not hold consistently. At best, it might be said that more of the industry's output is channeled into the production of capital goods. Steel in consumer goods as well as steel in foreign trade all affect this function. Beyond this is the fact that steel in ratio to capital stock generally (i.e., O_s/K) has increased historically as developments took place which made steel more plentiful and as technological developments throughout the economy resulted in production becoming increasingly more oriented to steel. As this has happened $\Delta GNP/\Delta t$ may have remained fairly constant, but it is found that over the same period steel output was not constant, rather it was increasing. In fact $\Delta O_s/\Delta t$ approached $\Delta GNP/\Delta t$.[7] However, in surveying the period 1884–1950, a tapering off of this growth in steel can be seen.[8] At present, it may be expected that the ratio O_s/K is more fixed

---

the relative capital intensity of evolving productive processes is not the only, nor necessarily a crucial, consideration in the construction of a dynamic economic environment. Fellner has treated this point as follows: "Significant changes in the composition of output tend to develop into general growth of the economy as a whole. The shift toward certain products stimulates investment more than is required to offset the under-maintenance in the fields adversely affected by the shifts. Consequently, tendencies toward changes in the composition of output may be important contributing factors of general expansion, which may extend even to the industries that would otherwise be adversely affected by the shift. This is one of the meanings that can be attributed to the dictum that the proper functioning of a capitalistic system required dynamic basic tendencies." See his *Monetary Policies and Full Employment,* pp. IX–X.

7. Cf. Table 1, notes f and g, p. 34.

8. Change in these series are reflected in Figure 1, p. 35. The 1884–1950 data appear to contradict what has been characterized as a period of capital intensification; in reality, it points to what appears to be a shift away from emphasis on capital intensification to what has been referred to as "improvement in the arts." For example, Duesenberry and Grosse state: "There are good reasons both empirical and theoretical for supposing that capitalist

or perhaps that output in the economy is even less steel-oriented in unit terms. If the latter is so, then $\Delta O_s/\Delta t$ should follow a trend of falling away from $\Delta \text{GNP}/\Delta t$. The relationship of a more nearly stationary steel output to sustain a rate of growth in the economy may then come to the fore. A condition is conceivable in these terms in which less steel output will be required in the future.[9]

In the Domar analysis, stability hinges on a very critical growth rate. If income levels advance too fast, insufficient capacity and perhaps insufficient savings accompany this growth. On the other hand, if income advances at too slow a rate, idle savings are to be expected and idle capacity becomes possible. Only at that critical stability rate will just the correct investment be taking place to utilize fully the savings generated, and at the same time will that very investment add to income an amount sufficient for the purchase of the expanded output.

What then can be said of the critical rate of growth in steel plant capacity? Expansion may be keyed to the stability rate of growth in the economy. Will it always be so? Factors might conceivably dictate a rate of expansion different from that of the economy at large. Is it legitimate to claim that the growth

---

economies of the type with which we are familiar do not have any tendency to adopt progressively more capital intensive techniques. Most of their progress stems from a process of throwing away old capital and replacing it with new without increasing the amount of capital employed per unit of output per year. Empirically there seems to be considerable evidence that on the average the ratio of capital stock to output per year has not increased over the last 75 years and may even have fallen somewhat. . . . Indeed it can be argued that under the kind of conditions which have ruled in capitalist countries there could be very little increase in output in the absence of change in the state of technology. There would still be plenty of room for substitution induced by price change which in turn could result from changes in risk, changes in the competitive structure of industry, changes in taxation or changes in wages brought about by trade union action or other causes. But those substitutions would not be systematic in character and might lower productivity as well as raise it." See A. Grosse and J. S. Duesenberry, "Technological Change and Dynamic Models," a paper prepared for the Input-Output Meeting of the Conference on Research in Income and Wealth (New York, National Bureau of Economic Research, October, 1952).

9. It should be mentioned that although steel forms less than ten per cent of output going into the investment total, characterization of the O_s/K relationship still has significance. Note that in this assertion there is implicit acknowledgment of the bottleneck theme.

rate in the industry will always be tied to that of the economy generally? In many past analyses of the nation's demand for steel, a fundamental premise has been that steel must keep "apace."

The formulation depends on the stability of the coefficients. In so far as the propensity to save and the aggregate marginal capital coefficient remain relatively stable, attention turns to endogenous coefficients, those of the capital coefficient within the industry and also the amount of steel involved in capital formation in the economy. If the assertion that the capital coefficient within the industry is relatively stable (at least, in the short run), then the coefficient which takes precedence is $\Delta O_s / \Delta I$. This prepares us for the discussion of demand analysis, which will be treated in Chapter 4.

Regarding the relative stability of the coefficients, a distinction should be noted between stability in the secular sense, as it was used in earlier remarks on capital-output ratios, and a condition of change in these relationships over shorter periods. The creation of excess capacity or, at the beginning of cyclical expansion, the replacement of capital in excess of depreciation, or possibly investment in processes with long gestation periods, may cause the actual capital-output ratio to exceed what might be the aggregative, technological, normative ratio. On the other hand, using up excess capacity at the beginning of expansion, or failure to replace equipment, or the ending of a gestation period, may create a situation wherein observed capital-output relationships fall short of the long-run norm. These generalized possibilities reflect the significance of a set of discrepancies very pertinent to the sequence just described.

In examining the series of $\Delta K / \Delta O$ and $C_s / \Delta O$ in Table 6 (page 59), the closeness of cyclical variation was seen. Only the increasing activity in the steel industry during the first half of the 1930s disturbed the year-to-year parallel in changes in the series. The formulation presented in Table 6, however, adhered strictly to an economic situation where output of the industry was tied exclusively to the production of capital goods and where the accelerator relationship was in the foreground.

Relaxation of the constraints on this picture seems appropriate, because in doing so the neat parallel movement of industry series and aggregate investment series tends to break down. Table 7 indicates a comparison between the $\Delta K/\Delta O$ and

TABLE 7. *Capital Stock Change–Steel Consumption Change Relationship, 1920–38*

Year	$\frac{\Delta K}{\Delta O}$	$\frac{\Delta C_s}{\Delta O}$
1920	5.4	5.4
1921	−1.5	11.3[a]
1922	.9	4.2
1923	.8	1.0
1924	5.5	−7.3
1925	4.0	3.9
1926	1.7	.6
1927	13.3	−6.3
1928	4.9	4.6
1929	1.7	.8
1930	− .6	2.3[a]
1931	− .02	1.4[a]
1932	.4[a]	.9[a]
1933	−4.5	9.4
1934	− .7	.5
1935	.1	2.6
1936	.6	1.6
1937	1.1	.1
1938	−1.8	13.1[a]

[a]Both numerator and denominator negative, yielding positive quotient, but nevertheless signifying decreases.

changes in the consumption of steel over changes in income ($\Delta C_s/\Delta O$) for the same period covered in Table 6. Here the years of discrepancy are more numerous, and appear to separate into two categories, one in which steel figures *exceed a normative figure* at the same time that aggregate $\Delta K/\Delta O$ *falls short* of a normative figure, and the other in which the reverse is true. The normative figure 3.5 was used in both cases, and arbitrarily chosen as being roughly the mean for the period covered. It can be seen that there is agreement in all years between the steel and investment figures in relationship to the 3.5 norm, with the exception of the following years: 1922 and 1933, when steel figures indicate a higher capital intensity than normative and where aggregate investment figures fall below the mark; 1924 and 1927 where the opposite indications hold. It might be suggested that in 1922, whereas there was some general employment of excess capacity in the economy at large, there was within the capital goods sector a specific program of replacement in excess of depreciation, and also possibly the suggestion of general investment in processes with a long gestation period. The year 1924 presents a picture wherein this situation is somewhat reversed, where the steel figures reflect general consumption of excess capacity, but where the aggregative picture is one of replacement. The year 1927 appears rather to be a reflection of the creation of excess capacity, but the steel ratio, which falls below the mark, reflects either failure to replace or the end of a gestation period. The 1933 discrepancy marks a manifestation of the beginnings of recovery in the capital goods sector, with a lagging response in the economy in general.

From the foregoing, it must be concluded that the problem is not merely one of equating growth in the industry with growth in the economy, either in the short run or in secular terms.[10] What is called for is an estimate of growth rates con-

10. Gordon states: "We cannot safely assume a constant total 'accelerator' either over the cycle or in the long-run; and, in fact, there is no particular rate of growth of total output which by and of itself, even if we could assume a constant marginal propensity to consume national income (or GNP), which would insure continued growth at the same rate. The process of growth itself requires changes in just those relations which current dynamic theories (such as those of Harrod and Hicks) assume to be constant. The cumulative process inherent in a private-enterprise economy tends to magnify these structural changes into the fluctuations we call business cycles. [Footnote] I have in mind here so-called major cycles. Reference to the cumulative process is intended to encompass financial and monetary factors, price-cost relations, changing expectations, as well as a flexible version of the multiplier process." R. A. Gordon, "Investment Opportunities in the United States before and after World War II," in *The Business Cycle in the Post-War World*, ed. Erik Lundberg, (London, Macmillan, 1955), p. 310.

sistent with the productivity of capital in the industry and in relation to the steel component of investment.[11] These factors will be considered in analyzing industry demand in Chapter 4. One aspect of the characterization of the role of iron and steel in the course of economic change still remains for discussion, however, before proceeding to the whole subject of quantitative demand analysis. It may prove to be the most crucial aspect of the industry's role, and therefore will be discussed in some detail.

LEVERAGE EFFECT AND BOTTLENECK PHENOMENA

In assessing the role of an industry such as iron and steel, in the structure of the national economy, it is possible to survey *ex post* output series for the industry and for the economy in general and to establish some kind of continuity, both in secular development and in cyclical fluctuation. This procedure, however, brings one merely to the point of saying that the output of steel has been a concomitant of aggregate change. It does not suggest sufficient evidence to establish causality in either direction, or if indeed there is causality in the first place. Does a significant relationship exist? Without going further than a

11. The considerable flexibility, and thus potential instability, of coefficients related to productivity is noted by Duesenberry and Grosse: "It is important to notice, however, that even a relatively advanced economy could increase its productivity at a higher rate for a considerable length of time. The age of the oldest capital in use in most industries is very considerable and the difference between the productivity with the oldest equipment and the best available equipment is rather large. It is therefore possible to speed up the rate of growth of output by a systematic reduction in the age of marginal equipment. While that process is going on, productivity would increase (a) because the productivity of workers with equipment of given age in successive years is improving (b) because the average age of equipment is being reduced. Such a process would, of course, require a very high rate of gross capital formation while it was going on. Moreover, one would not wish the process to go too far. If the life of equipment falls below a certain point, gross output would be expanded by a further fall, but net output would be contracted. It is probable, however, that even in the United States, the age of marginal equipment is higher than the optimum. This is obviously the case in many other economies." See "Technological Change and Dynamic Models," pp. 22–23. This subject is also treated by G. Terborgh in *Dynamic Equipment Policy* (New York, McGraw-Hill, 1949).

comparison of time series, we must be content with saying that there has been, in the past, steel output adequate for the general economic development undertaken— an *ex post* identity which is rather inconclusive. Can it be said in addition, however, that had there been more steel-producing capacity, Gross National Product would have been higher? Or conversely, that had there been less steel capacity, the economy would likely have experienced a lower Gross National Product?

These questions focus attention on two essentially different possible roles for the industry: what are treated here as leverage effect and bottleneck phenomena. In turning to the first of these questions, it is obvious that we cannot say that had there been more steel, there would have been higher national output. We recognize the multiple bases upon which investment decisions are made, and certainly such a statement would be a rash oversimplification of the forces at work. Nevertheless, we might consider, for the sake of thorough coverage, a perhaps rather extreme case of multiplier-accelerator interaction added to investment activity on the part of the steel industry. In such a case there is room, at least, for exploring further aspects of the question.

A claim has been made, for example, that if a capital goods industry, such as iron and steel, would only keep producing during the initial stages of a possible recession, such production could, in itself, prevent, forestall, or at least modify the downturn. This argument has been based on two premises. The first looks at "continuation of production" in terms of lowering prices to a point where the demand, otherwise shrinking, will revive and result in a stimulus to investment.[12] We will discuss aspects of this possibility in Chapter 6. The more direct implication of the phrase is what might be called the potential leverage effect of the industry.

Assume the economy to have reached the upper turning

12. Evidences of this approach are to be found in the final chapter of Machlup's study of basing points. Here attention is given to the possible effect of marginal cost pricing for steel during periods of falling demand so that demand trends might be reversed as a result of price concessions on the part of the industry. Cf. F. Machlup, *The Basing Point System* (Philadelphia, Blakiston, 1949).

point in the boom phase of a cycle. The year-to-year differences in national income have been growing smaller in each successive time period until they have approached zero. According to a simple formulation of the Hanson-Samuelson model,[13] at points where these differences have not as yet begun to shrink, accelerator action maintains its positive effect. At the point of inflexion, the accelerator loses strength.[14] At the upper turning point, induced investment disappears. Is it conceivable that autonomous investment could take place in such magnitude that it could halt or even reverse the negative effects of the accelerator?

In this formulation, the amount of induced investment is considered to be a function of changes in aggregate consumption. Therefore, for induced investment to be maintained at a given rate, consumption must grow at a rate sufficient to leave the differential unchanged; further, since consumption bears a functional relationship to income, the same requirement holds for growth in income. An additional dimension is added by introducing magnitudes of autonomous investment which either push up the base level of investment flow by a given amount because of the steady and continuous formation of this type of capital, or which at times effectively boost and at other times leave hanging in mid-air the other component of the investment sector.[15] For purposes of illustration, let us assume the possibility of compensating income-lowering deficiencies in in-

13. Cf. P. A. Samuelson, "The Interactions Between the Multiplier Analysis and the Principle of Acceleration," in *Readings in Business Cycle Theory* (Philadelphia, Blakiston, 1944), pp. 261–69.

14. Ragnar Frisch has called attention to the fact that if the rate of depreciation is high, the effect of a declining rate of consumption might be offset thus moving the critical point of contraction in capital goods away from the point of inflexion. Cf. his "The Interrelation Between Capital Production and Consumer-taking," *Journal of Political Economy, 39* (1931), 646–54.

15. The probable artificiality of this distinction between autonomous and induced investment has been pointed out effectively by D. H. Robertson, in "Thoughts on Meeting Some Important Persons," *Quarterly Journal of Economics, 68* (1954), 181–90, and by others. However, as with Hicks, the distinction is useful here to delineate that part of investment activity which might arise as a result of conscious policy decisions on the part of members of a particular sector of the economy, at a time when, perhaps, the "natural" conditions (expectations) favoring expansion in capital equipment might not so apparently dictate that course of action.

duced investment with investment decided upon independently.

The general principle this suggests is that the relative size of capital outlays in the steel industry, compared with those in other industries of a similarly integrated type, is significantly high[16] and that a capital expenditure in steel could be responsible for a counter-force to the pull of the accelerator at a particular point in time. This can be noted in the following.

Let K^* be the value of investment outlays attributable to steel plant expansion of a certain magnitude plus the effect of accompanying investment in agglomerating industries. K^* planning may have been instituted in a previous period, *but manifests itself at time* t.[17]

K investment induced as a result of changes in consumption (output) levels,[18]

Y income for a given period,

C consumption for a given period,

a accelerator relation,

ϕ the average propensity to consume,

t time period of one interval.

Then where $K_t = a(C_t - C_{t-1})$

and $C_t = \phi Y_{t-1}$

and $Y_t = C_t + K_t + K^*_t$

16. This fact is reflected in the capital-output ratios computed by the Harvard Economic Research Project, where the capital coefficient for steel works and rolling mills ranked tenth in the ninety-industry classification (a gross capital coefficient based on replacement cost of 1.798). Although justification of the focus of attention on the steel industry might be based solely on its recurrent prominence in discussions of public policy, it also might be said to stem from the nature of its product (combined with the high capital coefficient), i.e., a commodity providing a path to subsequent stages of expansion activity. This characteristic is less true of some of the even more capital-intense industries that precede iron and steel on the rank list; e.g., home renting, communications, trans-oceanic transportation, munitions, steam railroad transportation, transportation n.e.c.; more comparable claims might be made for others—petroleum and natural gas, manufactured gas, and public utilities. Cf. Leontief, *Studies in the Structure of the American Economy,* chap. 6.

17. K^* is an amount of investment representing capital goods any industry might purchase, but, in this case, action is taken exclusively by steel and agglomerating industries.

18. It may be assumed that the steel industry is operating at any given

the condition to be met is that

$$Y_t - Y_{t-1} \geqq Y_{t-1} - Y_{t-2}$$

K^*_t will be used to compensate for any "deficiency" in the income level of period t.[19] Within the assumptions of the model, it may be asserted that with given values of income in previous periods and an estimate of the accelerator and of the consumption coefficient, a "required" level of steel capital formation is represented by

$$K^*_t \geqq 2Y_{t-1} - Y_{t-2} - \phi\,[Y_{t-1} + a(Y_{t-1} - Y_{t-2})]^{20}$$

It can be seen that as the difference between Y_{t-1} and Y_{t-2} approaches zero, nearing the peak of the boom, the value of K^*_t would approach the value of the average propensity to save; i.e., savings would have to find outlet in autonomous investment. In years where the difference was larger than zero, K^*_t would have to be larger unless the accelerator was sufficiently strong to compensate for this. In any case, the dollar value of K^*_t would be of such magnitude as to make it beyond any reasonable steel outlay figure.[21]

Moreover, it is apparent that the value of K^*_t could merely forestall and not prevent the "inevitable" downturn, though it might conceivably ease the sharpness of the drop when it did come. Expenditures in successive periods or the institution of expenditures at an earlier date would fall in the same category; and if one attempts to find which continuing values of K^* might be sufficient to prevent the downturn, one is led along the paths of Harrod, Domar, and Fellner, to the equilibrium relationship

level of capacity, probably under 100 per cent. The rationality of undertaking a capital expansion expenditure at this time is not at issue for the moment. Expansion here is based on a decision "outside" of the framework of decision patterns in the industry; i.e., a policy decision in the context of national counter-cyclical policy rather than in terms of so-called long-run capacity needs and profitability. The paradox will be treated below in Chapter 6.

19. Note that because of the planning lag, it is assumed that "foresight" will be accurate enough to plan for K^*_t in advance.

20. By substitution from above and assuming K^*_{t-1} and K^*_{t-2} to have been zero.

21. Note that higher values for ϕ and a would ease the "task" for K^*_t.

between aggregate investment, the ability of new capital to increase output, and savings generated as a result of increases in output which are linked to a particular growth rate. The K^* factor might have significance if it were known to precede, suggest, or stimulate further flows of investment which perform the function of sustaining income levels. As we have considered the K^* variable to be additions to steel capacity, this line of reasoning suggests that we consider the possible agglomerating effects of steel expansion. The magnitude of these outlays may make the point.

An indication of the possible value of the variables can be seen in a post-World War II example of expansion of integrated capacity. A 1.8 million ingot ton plant was completed at the Morrisville, Pennsylvania works of United States Steel, including the facilities to produce semi-fabricated goods (bars, rails, shapes, etc.). The estimated cost of this addition to national steel capacity was $400 million. It was expected that additional steel capacity would be constructed in this area and that the ultimate total employment in steel production would reach 9,166 men. Employment in the plants themselves would be supplemented by an additional employment of 61,326 men in fabricating industries which might be expected to agglomerate as a result of the new capacity.[22] From this, a hypothetical

22. W. Isard and R. E. Kuenne estimated increases in employment to be expected within ten years with a total possible addition to steel capacity in the Delaware River Valley of 3 million ingot tons. The breakdown on their employment estimates indicated the form agglomerated employment takes: e.g., they estimated that total employment in the area including construction, government, agglomerated industry, and all feeder input industry employment which would arise just within the area studied, would come to 180,228 men. This they suggested was equivalent to a total population increase of 419,000 people. Cf. W. Isard and R. E. Kuenne, "The Impact of Steel Upon the Greater New York-Philadelphia Industrial Region: A Study in Agglomeration Projection," *Review of Economics and Statistics, 35* (1953), 294 ff. Subsequent reports on the agglomeration factor reveal that reality lagged expectations:

> Rumors of a new expansion program for the Fairless Works is one of the major interests in steel circles. Another is the active effort started recently by United States Steel to lure new businesses —steel consuming industries—to the Morrisville area. The number of businesses that followed Big Steel there—just two in five years—

figure for incomes generated by the Morrisville construction could be estimated to be approximately $260 million.[23] In the context of this discussion, however, the $400 million figure would be the significant one, if it is to be considered as a possible value of K^*.

One rather favorable aspect of this example is the relatively short interval between the actual institution of the expansion plan and its realization, a period of less than two years. On the other hand, the amount involved is less than 9 per cent of *monthly* national expenditures on new plant and equipment, based on monthly averages over the period 1947–50.[24] It is equal to approximately 1 per cent of the average annual gross private domestic investment over the same period. As for in-

---

has been disappointing to civic officials. In a bid to induce more consumers or suppliers to trail along, the corporation has put on the block almost one-third of the property it acquired. It has sent brochures offering thirty-one parcels covering 1,123 acres to some 300 of its customers. The corporation has said that 50 acres would be available immediately and the rest later on. It now has some 175 acres under roof and would retain for its own use a total of about 2,816 acres—enough for tremendously increased facilities.

. . .

In the Morrisville area the principal changes in the last five years have been: The borough's only bank has increased its assets from $6,530,000 to $11.7 million, has added one branch and is planning another. Real estate values have risen about one-third. The first sewer system, financed by a $3.5 million bond issue, is nearing completion. A stream of international commerce has started—each day there arrives an ore boat carrying an average of 7,000 tons of Venezuela's rich iron ore to feed iron and steel's principal ingredient to the United States Steel furnaces. Herbert Margerum, president of the Morrisville Bank, concedes the development in Morrisville has not proceeded as fast as civic officials had hoped five years ago. He adds, however: 'The mill has done a lot of good. When we get the sewer in we should be able to attract some small industry here and get badly needed apartment units. We have great hopes for revenue increases for business here.' "Fairless Works is Still Growing," *The New York Times* (March 4, 1956), p. F1.

23. Based on an average weekly wage of $70. The difficulty with using these figures stems from the difficulty in establishing how much of these earnings are transferred from other industries and other areas. E.g., Kuenne indicates that the net immigration inflow and natural increase in the area of his study would be 16.2 per cent. Effects of the deepening of capital per worker would have to be included in assessing the net effects.

24. Cf. U.S. Department of Commerce, *Business Statistics, 1951* (Washington, Government Printing Office, 1951), pp. 7 and 9.

come differentials, for example, quarterly figures for the years 1947–48 show two periods when the $400 million figure was better than 30 per cent of the $Y_t - Y_{t-1}$ and $Y_t - Y_{t-2}$ differential.[25] It seems doubtful, however, that this presents grounds for optimism—since the quarterly changes represent time intervals virtually "unmanageable" in policy sense. The use of differentials based on annual totals for income dwarfs the steel figure significantly.

It must also be recognized that the effect of this capital outlay can be expected to be relatively short-lived, and only a stopgap in the most transitory sense. This is true even if K^* were maintained over successive time periods. If it were possible to estimate the size of the outlay necessary to forestall the turning point, it is still highly improbable that a meaningful quantitative estimate can be tied in with the investment planning and timing pattern. Three main conditions would have to be met: (1) the upper turning point of the cycle would have to be anticipated in advance by an interval which would coincide with the planning-period to fruition-point interval of investment in the industry; (2) this point would have to be predicted with sufficient accuracy so that the magnitude of income differentials would be relatively small. To be more precise, the magnitudes would have to be small enough to bear some comparative relationship to reasonable capital expenditures in the industry plus the accompanying agglomerative outlays; (3) the size of capital expenditures in the industry would have to be such that their addition to the consumption occurring in the economy (plus whatever multiplier effects could manifest themselves in the time period) would result in an increase in national income over the previous period greater than the corresponding increase in that period over the one preceding it. All this, it must be remembered, must occur at the point in time when, without the industry investment expenditure, income differentials would have fallen. Even in the absence of

25. Ibid., p. 6. In January 1949 the reversal set in and continued through that year. Note that in any quarter, steel plant expenditure would be a fraction of the $400 million figure.

these requirements, it would be unreasonable to conceive of one industry carrying the leverage load over an extended period.

What, then, may we conclude from all of this? We have covered the ground because of assertions that capital expenditures in the iron and steel industry might be expected to have sufficient leverage effect to help forestall the turning point in the boom. Our discussion has served to qualify, if not to dismiss, such statements. They are true under the most rarefied and critically timed conditions, when all magnitudes are such that the impact of the additional autonomous investment (in a given period or over subsequent periods) heads off the downward turning of the peak. But, as we have seen, timing would have to be such that relatively small magnitudes of $Y_t - Y_{t-1}$ and $Y_{t-1} - Y_{t-2}$ would coincide with outlay on new plant and equipment of the industry. At best, we may say that capital outlays in steel at a time of prosperity *might* play an important part in a *larger* program to affect the pull of the accelerator.[26]

In spite of this pessimistic judgment, one additional point may be added. Our attention has concentrated on what might be termed *real* factors affecting the cycle, that is, investment and its impact on changing levels of income. There is another possible consideration. If, for example, the prosperity phase of the cycle can be said to be losing its force because of a readjustment in the expectations of the entrepreneur, or of the banker, or even of the consumer, then it might also be said that op-

26. Leverage effects in the depth of a depression are a possibility not considered in the material above. Reference to the magnitudes of capital formation and income levels at a low point in the cycle suggests that an expenditure such as Morrisville would have much greater relative effect at such a time than in the prosperity phase. The writer has refrained from discussing this aspect of the cycle not only because the inquiry has been directed toward an evaluation of contentions regarding the industry during periods of relatively high employment, but also because motivation to expand within the industry at other times appears limited. There are specific examples where expansion was undertaken during the low point in the 'thirties, but this type of programming does not appear consistent with what seems to be a more valid generalization of investment behavior, that is, in response to a favorable demand picture. In addition, there is the likelihood of large excesses of capacity at just such times. These possible excess conditions call for special consideration and, of course, have occurred during periods of full employment. This will be discussed in Chapter 6.

timism exhibited by one sector of the economy might well change the picture. If individuals, or firms, or an industry, were to take decisive action just at the time when investment opportunities seemed to be drying up, and if these individuals were respected and recognized by the public at large, a change in the tide of the cycle might well occur.

Assume the economy to be approaching the upper turning point of the boom. This point occurs, in part at least, as a result of a change in expectations. Among the circumstances that affect expectations are changes in the price level, the rate of interest, the flow of invention, and the many other factors in cycle analysis. There is also the additional phenomenon sometimes termed "confidence." Confidence is a difficult variable to pin down, no doubt, because of its tie to a range of subjectively determined elements. Nevertheless, assume that certain events will have greater impact than others on the confidence felt by decision-makers, and let us assume that the impact of leaders in the iron and steel industry on others is large—larger than that of many other leaders in the business community. The steel industry does have a "halo-effect" operating in its favor which may allow us to attribute this role to it. For example, a casual introductory remark by Senator Walter O'Mahoney at a Senate hearing conveys this feeling:

> The significance of a hearing on steel is not that the Government has a large investment in steel plants when compared with the investment in other industries, so far as percentages are concerned, but it lies in the fact that the production of iron and steel throughout our recent economic history—at least in the last half century—has been the measure of our prosperity. When there is a high rate of steel production, times are good; when there is a low rate of steel production, times are bad.[27]

27. U.S. Congress, Senate, "War Plant Disposal—Iron and Steel Plants," *Joint Hearing before the Subcommittee on Surplus Property of the Committee on Military Affairs and the Industrial Reorganization Subcommittee of the Special Committee on Postwar Economic Policy and Planning,* 79th Congress, 1st Session, November, 1945 (Washington, Government Printing Office, 1946), p. 3.

Moreover, the combination of business community respect for the industry's leadership with the importance popularly attributed to the industry in a fluctuating economy makes the steel industry's impact on community confidence enormous and uniquely so.

If the upper turning point of the cycle is the result, in part at least, of a diminishing confidence felt by the entrepreneur and the banker, and if at the same time the confidence of these individuals is highly responsive to certain generalized actions in the community—for example, to what they may see happening in the "central" and "dominant" industry, steel—we may have another form of "leverage" to suggest. If, as the upper turning point in the boom is approached, leaders in the steel industry act[28] to continue current rates of output, to continue inventories at the current level, or to expand plant facilities, the impact on the confidence level of the community at large may be very forceful indeed.[29]

In other words, action on the part of the industry might have the effect of shifting to the right the marginal efficiency of capital schedule purely on the grounds that general expectations rise as a result of the change in the direction of confidence. In causing a reversal of a declining state of confidence, the action of the steel industry would have far greater impact than in the illustration above, where, as an item of autonomous investment attempting to stem the tide of declining induced investment, it proved too weak for the task. With a shift of the marginal efficiency schedule, however, the whole range of incentives to invest would be revised, and even with an advancing rate of interest or a considerable amount of capital already formed, there still would be stimulus to invest further. This generalized action might be able to reverse the course of what would have

28. It must be recalled again at this point, that speculation in this part of the study is on the possible effect of industry action on the economy; whether it is reasonable to expect the industry to take such action is not at issue here.

29. This is an area of discussion which suggests a point of contact between the expectation factors of Keynes and the entrepreneurial reaction patterns of Schumpeter. Cf. J. M. Keynes, *The General Theory of Employment, Interest and Money* (New York, Harcourt, Brace, 1936), pp. 143–44 and pp. 315 ff., and Schumpeter, *The Theory of Economic Development,* p. 230, on expectations and leadership respectively.

been a recession, whereas the small autonomous investment of the steel industry alone, when considered in real terms, was unable to do the job.

The preceding argument serves to establish the tenuous nature of the supposed "positive role" of the iron and steel industry. It must be acknowledged that increases in the industry alone can do little to prevent a recession. The most that can be said is that the earlier in the recovery or prosperity phase capital outlays are made in the steel industry, as in all others, the better the chance of picking up momentum through the supermultiplier. Moreover, though it was suggested that by taking certain psychological parameters into consideration, a hypothetical situation can be constructed where the industry might affect confidence, here we are led to very limited conclusions based on tenuous assumptions. It appears that this is not the area in which steel plays its most crucial role. In other words, the industry's role as an autonomous element is negligible; alone, it has little potential multiplier effect. On the other hand, as shall be seen in what follows, the industry's potential as an inhibiting factor in the course of economic change is significant.

An early reference to "bottlenecks" specifically in relation to the upper turning point of the cycle was made by Haberler, in writing of an interruption in the expansion process:

> If the expansion does not happen to be directed along "the path of least resistance" as determined by the distribution over various industries of the available resources and their mobility, the situation which we have discussed will arise before all unemployed factors have been absorbed. This will manifest itself in the emergence of "bottle-necks"—that is, in the increasing scarcity of some factors of production leading to a rise in prices of certain commodities and a slowing-down of the expansion in this or that industry.[30]

30. G. Haberler, *Prosperity and Depression* (3d ed. New York, United Nations, 1946), p. 371.

He continues by pointing out that these bottlenecks may appear in one or several sections of the economy, depending in some measure on the given distribution of resources, and therefore of capacity, at the beginning of the upswing. In the case of an expansion dominated by high investment activity, it follows that the limits to expansion may well be reached in one or several of the capital goods industries.[31] In Haberler's account of the sequence of events, the bottleneck is reached before *all* unemployed factors are drawn into use—focusing attention on the fact that the so-called downturn from the ceiling characteristic of the Hicks and Goodwin models[32] may well be in reality a sequence of events following very selective scarcity—scarcity which cannot be relieved through substitution.[33]

Fellner, writing on this point, first acknowledges that significant scarcities may not be apparent in certain downturns, but then goes on to say, "However, certain phenomena in the *nature* of scarcities could probably be demonstrated for any upper turning point one might select to study,"[34] and further, "it is not necessary to make extreme assumptions with respect to the existing degree of scarcity in order to use a pattern resembling that of the overinvestment theories (in conjunction with other patterns)."[35] This latter because, as Haberler pointed out, "the economy may become gradually more sensitive to random disturbances as its resources become more fully utilized."

To localize these references to the role of the steel industry: though it would be difficult to assert that specific additions to

31. Or for that matter by one segment in a given industry, e.g., pig iron production—shortages of which, when they coincide with limits to scrap supply, can set limits to significantly usable finishing mill capacity.

32. Hicks, *A Contribution to the Theory of the Trade Cycle;* R. M. Goodwin, "The Nonlinear Accelerator and the Persistence of Business Cycles," *Econometrica, 19* (1951), 1–17.

33. This selective scarcity might well co-exist with overproduction in other sectors, but for our purposes, it is to the former's impact that attention will be given.

34. W. J. Fellner, "Employment Theory and Business Cycles," in *Survey of Contemporary Economics*, ed. H. S. Ellis (Philadelphia, Blakiston, 1948), pp. 65–66.

35. As Robertson has observed, in the sense used here the bottleneck ap-

capacity in the industry would act to prevent a downturn, there seems to be much more force in the claim that a limit in steel capacity might play a significant part in bringing on the downturn.[36]

In the context of the generalized theory of over-investment (or selective under-investment), the steel industry might be the guilty party. The specific bases on which government demands for additional steel capacity[37] were made in late 1947 and 1948 were so-called projections of future steel consumption patterns.[38] The contention was that possible shortages in steel could choke off continued output in various other sectors, precipitating cutbacks all along the line. The relative merits of the estimates presented at the time are, in one sense, not subject to *ex post* evaluation because the Korean emergency arose after

---

paratus seems to lead to what might more appropriately be termed an under-investment theory, or a selective under-investment theory. Cf. D. H. Robertson, "A Revolutionist's Handbook," *Quarterly Journal of Economics, 64* (1950), 2, n.2.

36. The "size" of the cycle is not at issue here, though whether this scarcity might precipitate a major downswing seems to depend on which other factors and forces are operative. As Duesenberry indicates: "Ceilings on resources may help to explain short minor cycles, but it does not seem very likely that resource limitations *in themselves* will cause major depressions. Nonetheless, resource limitations may play an important role in bringing on major depressions. They may do so in a number of ways. (1) A reduction in the rate of growth of income may be the proximate cause of the collapse of a speculative boom. (2) A labor shortage may cause a wage-price spiral and a general speculative boom, and a capital-goods shortage may bring on a boom in the capital-goods industries. In 1873, rising prices for coal and iron contributed to the collapse of the American railway boom and also caused a boom (which later collapsed) in the development of British coal mines. (3) Shortages of labor or of funds may reduce the flexibility of the economy so that expansion in one sector cannot offset contraction in another. . . . Ceilings on resources are not in themselves likely to produce major depressions because they do not produce large reductions in the rate of investment and because a temporary reduction in the rate of growth of income is sufficient to remove their influence." J. S. Duesenberry, *Business Cycles and Economic Growth* (New York, McGraw-Hill, 1958), pp. 280–81.

37. Cf. above, page 16, note 1.

38. A notable tie-in of needs in steel capacity, for example, with the maintenance of full employment was found in the study by J. Cornfield, W. D. Evans and M. Hoffenberg, at that time. Cf. "Full Employment Patterns 1950," *Monthly Labor Review, 64* (1947), 163–90. Evans and Hoffenberg have indicated disposition of steel output in "The Inter-industry Relations Study for 1947," *The Review of Economics and Statistics, 34* (1952), 122–24.

the calculations were presented.[39] Because of the Korean impetus to the demand for hard goods, estimates of a need of 110 million ingot tons proved to be close to what was "adequate" in 1950. Industry representatives, however, considered this figure to be unreasonably excessive in 1947–48.[40]

To cast this discussion more directly in the context of recent models of cyclical change,[41] the following material will present a construction wherein bottleneck conditions are introduced. It is, of course, beyond the scope of this study to evaluate the literature on the acceleration principle. However, a qualification of its usage is called for, in the manner that several writers have emphasized in their evaluation of the principle.[42] The accelerator will be used below in a model which permits its variability over the cycle, and which is not dependent on its being strong enough to combine with other factors and produce an explosive cycle, although this might happen.[43] The phenomenon of investment in response to changes in income is assumed; but so is the feedback effect on income as a result of new capital formation. The time interval of the cycle is variable in the sense that it is expected to change dimension with a change in one or several parameters.[44] Whether the response in investment activity has worked via changes in price, or in profit relations, or more directly, is not a confining consideration.[45] Finally, in the

39. Some take the position that it is precisely this inability to incorporate what may turn out to be profoundly significant data, that vitiates much of the usefulness of the type of projection undertaken.

40. U.S. Congress, Senate, *Hearings Before the Special Committee to Study Problems of American Small Business,* 80th Congress, part 8, "Steel Supply and Distribution Problems Affecting Smaller Manufacturers and Users," Testimony of Wilfred Sykes, June 19–20, 1947 (Washington, Government Printing Office, 1947), pp. 984 ff.

41. Those of Hicks (*Contribution to the Theory of the Trade Cycle*), and Goodwin, ("The Nonlinear Accelerator").

42. E.g., S. S. Alexander, "Issues of Business Cycle Theory," p. 877.

43. In this connection Fels has suggested that an independently "weak" accelerator may have effective strength because of concomitant innovation. He writes of a synthesis of the Hicksian and Schumpeterian approach. Cf. R. Fels, "Theory of Business Cycles."

44. The relevant contrast, for purposes here, would be in the distinction between the period of the "uninterrupted" cycle and one in which the choke-off process appeared.

45. Cf. Alexander, "Issues," pp. 868–69.

following formulation the question of the specific level of income and the investment needs resulting therefrom are combined with needs resulting from changes in that level of income. All of these qualifications hold, in this particular case, because the sole objective is to set up a cyclical sequence and then to impose on it the condition of specific scarcity in some area of the capital goods sector.

Without repeating the development of the models of Hicks and Goodwin, in which the possibilities of completely internally contained cyclical mechanisms are presented,[46] it should be noted that the cycle does not depend on reversals resulting from the response to price changes in capital good markets,[47] but that just as conveniently, "the whole process can be translated into terms of investment resulting from a discrepancy between desired and actual capital stock (with a rigid supply price for new capital goods throughout)."[48] For both writers, the imposition of a restraint not only results in the non-linearity of the cyclical mechanism, but it also sets the conditions which prevent the system from flying apart. In Goodwin's model, for example, the supermultiplier might well be so large that the feedback prop-

46. Hicks places more emphasis on limits to expansion because of the full employment ceiling; Goodwin, on the other hand, devotes more attention to the limiting factor of full capacity operations, particularly in the capital goods sector.

47. It might be said that one of the weaknesses of Goodwin's model is that it ignores the effects of price changes of capital goods. However, for our purposes this is a most appropriate structure. The steel producer does not (in the short run) follow "clinical" profit-maximizing behavior; the oligopolistic structure of the industry plus the Kefauver effect results in the Goodwin constraint being applicable.

48. R. M. Goodwin, "Econometrics in Business-Cycle Analysis," in A. H. Hanson, *Business Cycles and National Income,* (New York, Norton, 1951), p. 453. Tobin points to a general weakness in this type of model:

> Goodwin's ingenuity makes it possible to go a long way in describing the twin processes of growth and cycles in terms of aggregate capital requirements alone. But not far enough. . . . An accelerator model, whether flexible or rigid, cannot explain inflation and deflation, nor can it allow the lengths of booms and depressions to depend on fiscal and monetary events and policies. Goodwin's model illuminates half of an important truth: whether the economy is booming or depressed depends on whether capital is short or redundant. It misses the equally important other half of this truth: the adequacy or redundancy of the stock of capital at a given in-

erties of the system would drive the swings further and further from equilibrium, were it not for the ceiling imposed by the discrepancy between desired and realized plant capacity. The implications, then, are that the possibility of a choking action of a given magnitude of capacity, in steel for example, might well be a contributing factor to the relatively moderate proportions of cyclical fluctuation. (This, in addition to the case where a bottleneck condition may act to constrain or shorten an already damped cycle.)

In the accompanying diagram (Figure 6), the curves represent the demand for or valuation of investment goods at differ-

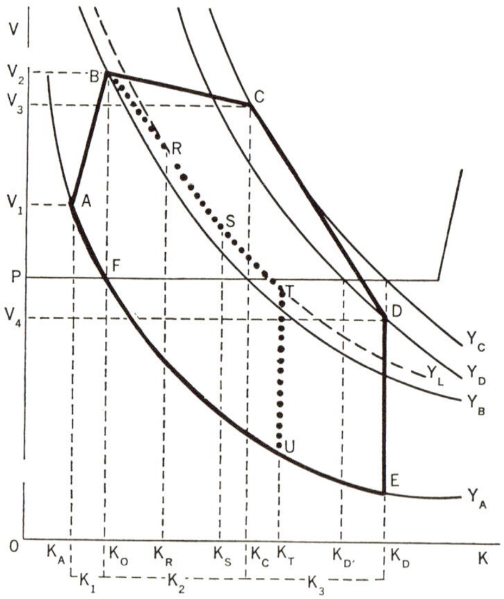

Figure 6. Investment Demand and Income: Model of a Constrained Cycle

---

come level is not a knife-edge question, but depends on the relative supplies and yields of capital and alternative means of holding wealth.

J. Tobin, "The Business Cycle in the Post-War World: A Review," *The Quarterly Journal of Economics*, *73* (1958), 290–91. It is useful for our purposes, however, to illustrate the potential action of that sector of the economy under scrutiny here.

ent levels of income, so that they reflect demand as a function of the amount of capital stock and the level of income, and also of the force of the accelerator and the state of expectations in the slope and shape of the curves. The multiplier works to affect the differential gap between the various levels of the Y curves. It is assumed that the multiplier effects are greater at initial injections of new investment and taper off as K (or $\dot{K}$) increases —that is, the elasticity of a curve plotting the differential gaps falls with increases in $\dot{K}$.[49] In this example, the price of capital goods is assumed to be quite rigid in the significant range, at least over the relevant period. If it is assumed that desired capital stock (K) is the result of income level, actual stock, and price, and that investment takes place in response to the gap between actual stock and desired stock, then the sequence over the cycle can be shown.

As a starting point, it may be assumed that more than adequate capital stock exists and that negative net investment dominates the picture. Income level would be Y_A, and this curve represents the demand for capital goods at this time. Movement will take place along this curve, from for example E in the direction of F as unused capital stock becomes fully depreciated and replacement needs increase. At some time, however, a point on Y_A beyond F will be reached,[50] e.g. point A. Here the valuation of capital goods would be V_1 and in excess of the price P. Here desired stock would be in excess of the actual by the amount K_1. If it follows that investment takes place in the amount $\dot{K}_1$, then because of the multiplier effect ($k\dot{K}$) addition of this amount to stock will mean that, although total stock is now equal to OK_o, movement is on a new demand curve Y_B ($= k\dot{K}_1$)and valuation would be V_2. Desired stock would be OK_c (because, again, $V > P$), and investment $\dot{K}_2$ would take place. This would result in actual stock of OK_c, but also move-

49. See Goodwin's plot of this in his ϕ_2* curve, in "Econometrics in Business Cycle Analysis," p. 451.

50. At any distance above F, that is. F is not an effective point of equilibrium as it would be only a matter of a slight lag in the flow of capital goods (a backlog of orders) that could create some distance FA where desired stock would exceed actual stock.

ment to curve Y_c ($= k\dot{K}_2$). Once more there is an excess of desired over actual stock and $\dot{K}_3$ is undertaken. $\dot{K}_3$, however, is less than $\dot{K}_2$ and therefore creates a new demand situation at the level of Y_D ($= k\dot{K}_3$). Desired stock is now $OK_{D'}$, but the actual supply has been increased to OK_D resulting in an excess of actual over desired (reflected in V_4 being below P).[51] No further investment takes place; rather there is disinvestment and this produces the sharp drop back to the Y_A curve and point E, beginning the process once more ($EABCDE$). Note that so long as the multiplier effect on income raises the demand curve more than the gap between desired and actual stock, there will be an upward swing; when they reverse, K will be catching up. At some point the demand for K will have fallen so low as to pull the Y level down, completing the cycle.

The next step is the introduction of a case of specific scarcity, for our purposes, that of steel capacity, and the tracing of its effects on the cyclical pattern. A parenthetical reference to what was termed "selective under-investment" is appropriate here, for it might be thought that this construction is one in which lack of investment response from within the industry is necessary. This is not the case, for, in line with Hicks' reasoning, the incentive to invest may be operative, and steps may be taken by steel management in this direction; however, if there is a significant lag, the constraining action will still affect industries using steel inputs, thus causing cutbacks in their output.[52] Once this occurs, a situation of selective scarcity can exist: for example, output in consumption goods may still be rising, but the stoppage in the flow of investment goods may result in severe retardation in the rate of expansion; for at this time, it is only via deferred multiplier effects that income levels are stimulated, the force of the accelerator having been withdrawn.[53]

If the limitation of a specific scarcity is imposed on the cycle

51. The interval necessary to reach this point of excess might be more prolonged, and the movement from B to C to D a rather gradual one.

52. Cf. Hicks, *Contribution to the Theory of the Trade Cycle,* p. 99 n. 1, and Duesenberry, "Hicks on the Trade Cycle," passim.

53. Hicks, *Contribution,* pp. 128–29.

sequence discussed above, it can be shown that a limit appears to the level of capital demand curves, and that the time intervals necessary to reach a given amount of capital stock increase in number. Assume that the maximum amount of capital stock formation that can take place in one time interval is confined to an amount equal to K_oK_R (on Figure 6 above). This means that, although at point B desired stock exceeds actual stock by the amount K_oK_c, only the amount K_oK_R can be supplied; and this means that rather than a demand curve boost to Y_C, movement is to curve Y_L ($= k\dot{K}_R$)—that is, to point R. As demand for stock still exceeds supply, investment will continue to take place, but each time limited in amount equal to K_oK_R. Additions are made subsequently to K_S and K_T (all along the Y_L curve) until finally actual stock exceeds desired stock and, as in the case above where there was a drop from D to E, there is a drop from T to U. The cycle sequence now has become *UABRSTU*. In this context it can be said that the nature and "severity" of a bottleneck condition can key the dimensions of a given cycle, specifically in the sense of imposing a constraint on the cycle. If the limit were increased sufficiently, to cite an extreme, the economy would in effect be backed into a stationary box with no dynamic potentialities. A more reasonable suggested magnitude might lead to the conclusion that a bottleneck condition could precipitate a downturn sooner than would have been expected, or conversely could shorten the period of expansion in that phase of the cycle.

Although the model has been presented more in terms of stating the range of possibilities associated with a bottleneck effect than of asserting its relative likelihood, the latter must also be treated. This will be done in Chapter 6. However, the foregoing argument indicates what is implied in the assertion of the bottleneck potential of an industry such as steel. The theoretical formulation and its implications for policy formation are specifically relevant so long as the possibility of so-called selective under-investment and scarcity is present.[54]

54. The "importance" of this sequence rises, of course, as a concomitant of inadequate levels of aggregate autonomous investment. Hamberg writes of

Attention to the question of measurement of demand should precede further discussion of the bottleneck possibility. If an adequacy measure for iron and steel capacity were readily available, then the industry's potential role as a choke-off to expansion could be appraised more accurately. Further, there would be ground for more general agreement on the scope and direction of programming which the industry might best consider.

"horizontal maladjustments," by which he means "inelastic expectations, sectoral shortages, and surpluses"; he states: "When autonomous spending outlets (including government spending and, significantly, residential constructions—as the events of recent years have shown) are great enough for investment from these sources to provide strength and resiliency to the aggregate economy, the latter is not overly sensitive to horizontal maladjustments. . . . On the other hand, when autonomous investment outlays in sufficient magnitude are lacking, these very same maladjustments may (though of course not necessarily) be the last straw." D. Hamberg, *Economic Growth and Instability: A Study in the Problem of Capital Accumulation, Employment, and the Business Cycle* (New York, Norton, 1956), p. 309.

FOUR QUANTITATIVE ANALYSIS OF DEMAND FOR INDUSTRY OUTPUT

PREVIOUS ATTEMPTS TO ANALYZE DEMAND FOR INDUSTRY OUTPUT

A prewar study of possible future demand for steel was made by the National Resources Committee in 1938.[1] In their publication, *Patterns of Resource Use,* an attempt was made to estimate future demand for various products on the basis of possible alternative future levels of consumer income.[2] They assumed that for each level of income there was a corresponding level of industrial production, and that for each of the levels of industrial production there were corresponding levels of demand for products. The same logic was applied to employment in each industry and for the economy as a whole. For example, actual values for consumer income (in billions of 1936 dollars) for the years 1929 and 1935 were 65.3 and 56.9 respectively. Corresponding to these were indices of industrial production (1923–25 = 100) of 119 and 90 respectively. And further,

1. Earlier demand estimates for steel have been surveyed by Paul Boschan in "Productive Capacity, Industrial Production, and Steel Requirements," in *Long Range Economic Projection,* National Bureau of Economic Research (Princeton, Princeton University Press, 1954), and includes reference to: G. C. Evans, "The Dynamics of Monopoly," *American Mathematical Monthly, 31* (1924), 77–83; C. F. Roos, "Dynamical Theory of Economics," *Journal of Political Economy, 35* (1927), 632–56; R. H. Whitman, "The Statistical Law of Demand for a Producers' Good as Illustrated by the Demand for Steel," *Econometrica, 4* (1936), 138–52; and United States Steel Corporation, TNEC Paper, pamphlet no. 5, "A Statistical Analysis of the Demand for Steel, 1919–1938" by H. L. Gress and T. O. Yntema.

2. U.S. National Resources Committee, *Patterns of Resource Use* (Washington, Government Printing Office, 1938), p. 27.

corresponding production figures for iron and steel (in millions of gross tons) were 51.3 and 31.7.

TABLE 8. *Patterns of Resource Use—Iron and Steel*[a]

	Actual values for		*Assumed values*					
	1929	*1935*						
Consumer income (billions of 1936 dollars)	65.3	56.9	50	60	70	80	90	100
Corresponding industrial production (1923–5 = 100)	119	90	68	90	113	139	166	195
	Actual patterns of Resource use for		*Calculated patterns of Resource use assuming 1938 trend values*					
	1929	*1935*						
Iron and steel (millions of gross tons)	51.3	31.7	15.7	31.7	48.5	67.4	87.1	108.2

[a]U.S. National Resources Committee.

Resource use patterns were derived from actual patterns of resource use over time, based on trend projection formulas, using assumed values of consumer income and of corresponding industrial production. Table 8 is abstracted from the complete set, which covers 81 segments of the economy. In 1941, consumer income approximated the 90 billion dollar value (actually it was closer to 94 billions) which would suggest that steel production should be 87.1 gross tons or 97.6 short tons. It was in reality 82.8 short tons, with the industry operating at 97.3 per cent of capacity.

The National Resources Committee completed a further study, based on the *Patterns* estimate, with the intention of setting forth capital requirements in the industry at alternative levels of income.[3] The logic proceeded along the following lines: steel production was defined as the sum of domestic steel consumption and net exports of steel; steel consumption was obtained from the various estimates at alternative levels of income given in the *Patterns* study described above; net exports were derived by obtaining the difference between estimated exports and imports of iron and steel products. The steel production figures thus obtained were then used as targets for breakdown estimates of raw material and specific production facility requirements. From these requirement figures, existing capacity was subtracted, and needed additions to plant were calculated. The particular contribution made by this study was in the attempt to distinguish the input and finishing capacities by specific type. Raw material relationships, for example, were treated item by item: scrap consumed in steel-making and its relation to steel ingot production, relation of pig iron to ingot production, production of ferro-alloys, iron ore consumption and pig production, coke consumed in blast furnaces and pig produced, limestone and scrap consumed were all included. This was also true of production relationships; capacity requirements for blast furnaces, steel works, and finishing mills were all treated separately. As we shall see, this is a particularly advantageous approach.

In the early part of 1941, President Roosevelt asked Gano Dunn, an engineer attached to the Production Division of the Office of Production Management, to make a study of the adequacy of the steel industry in terms of national defense needs. Dunn's report was significant for several reasons.[4] It set out to estimate industry capacity in a manner keyed to the technological structure of the industry. It also attempted to integrate estimates of civilian demand for steel at various levels of national income with the needs of an expanding armament pro-

3. U.S. National Resources Committee, *Capital Requirements, A Study in Methods as Applied to the Iron and Steel Industry* (Washington, Government Printing Office, 1940).

4. Dunn, "Report to the President of the United States on the Adequacy of the Steel Industry for National Defense," (Washington, Office of Production Management, February 22, 1941).

gram. It will be useful to examine this study rather closely.

Before doing so, however, a brief note on the use of capacity figures in the industry is needed. In the material which follows, there appears a somewhat anomalous statistic giving capacity utilization as greater than 100 per cent. Capacity figures for both blast furnaces and steel furnaces are calculated on past performance, including allowances for periods when the furnaces are idle because of regular maintenance shutdown. In the case of blast furnaces, for example, not only are the specific characteristics of a given furnace and its physical dimensions taken into consideration, but also the particular materials used to charge that particular furnace. In some cases, furnaces are operated with high iron-content ores and in other cases with ores having as low as 35 per cent metallic content.[5] In addition to consideration of ore quality, the characteristics of coke and other materials are included in the calculation of the furnace's capacity. In calculating the capacity of a particular furnace, past experience is the basis for this figure. The amount of production under full operating conditions is determined, and then an annual "rated capacity" is derived by extending this production figure for the actual operating time expected, for a period of one year. An example of the capacity figures derived by this method is seen in Table 9. A similar technique is used for determining the ingot capacity of a given open hearth plant.

> The annual capacity is computed by multiplying the number of days in the current year by twenty-four hours per day by the number of furnaces less the number of hours the furnaces will be idle due to lost time on account of relining and rebuilding and due to holiday shutdowns, times the expected net tons per operating hour. However, lost time does not include time for actual fettling, repairs to bottoms, banks and tap holes, hot work repairs (roof patches, replacement of front walls, etc.), mechanical repairs and other operating delays. Capacities for bessemer

5. This information is taken from "Steel Capacity Statistics Based on Performance," *Steel Facts, 113* (1952), 6.

TABLE 9. *Method for Computing Blast Furnace Capacity of a Representative Plant*[a]

	Size and working volume			*Annual production record*			*Annual capacity January 1, 1952*				
								Less annual deduction for rebuilding, repairs[b]			
Furnace number	*Height (feet and inches)*	*Hearth diameter (feet and inches)*	*Working volume (cu. ft.)*	*Year*	*Net tons*	*Type of metal*	*Gross annual capacity, net tons*	*Per cent*	*Net tons*	*Per year*[c]	*Net capacity: net tons Per day in operation*[d]
A	103′–8″	28′–0″	50,336	1951	495,783	Pig iron	559,500	11.3	63,100	496,400	1,533
B	96′–6″	20′–0″	30,500	1950	314,304	Pig iron	332,632	5.0	16,632	316,000	911
C	94′–2″	23′–6″	36,735	1950	336,494	Pig iron	354,800	5.5	19,500	335,300	972
D	85′–3″	18′–6″	23,098	1950	84,966	Ferroalloys	90,200	5.5	5,000	85,200	247

[a]*Steel Facts*, p. 6.

[b]Includes estimated per cent of time and net tons lost when furnaces will not be in blast due to rebuilding, relining and repairs. Any minor delays such as changing tuyères, etc., are not included as deductions.

[c]Represents difference between Gross Capacity and Deductions.

[d]Days in operation represent 365 days less the number of days estimated to be lost for deductions for relining, rebuilding, and repairs.

converters, electric and crucible furnaces are computed similarly.[6]

At the outset of his report, Dunn notes that definition of capacity figures has a definite bearing on maximum possible output calculations. He emphasizes that published industry figures are "rated capacity" figures. They were derived by subtracting from technologically determined capacity a ten per cent allowance for maintenance and repair shut downs.[7] The importance of this, as Dunn interpreted it, was that the ten per cent leeway was unduly generous when the industry was faced with a war emergency. By incurring a slightly higher cost, it seemed likely that the shut-down period could be shortened yielding a figure for "reliable capacity" of 102½ per cent of rated capacity. 105 per cent could probably be obtained under considerable pressure.[8] It is in this sense that the reliable capacity figures discussed here should be understood.

In examining the nature of capacity increases in the industry, Dunn pointed out that it was a characteristic of the industry that capacity increases typically occured when action was taken to reduce costs, i.e., that the additional capacity was, in a sense, a by-product. The emphasis in employing new technology or a new location was on cost minimization or quality improvement. He noted that during the depression, 1930–39, capacity increased at the same time as output fell to an average of only 49 per cent of capacity. He made the point that new capacity does not displace old capacity. Rather, the old capacity is used for stand-by purposes, in case of emergencies, or operated at a reduced rate to produce some kind of special steel.

The shift to more efficient methods of production was reflected also in the shift from one process to another. For example, in 1907 Bessemer steel production amounted to 13,067,655 short tons, open hearth to 12,935,704 short tons, and

6. Ibid.

7. E.g., over a year's time, approximately five weeks are expected for shutdown, to reline furnace walls and so forth.

8. E.g., in the *Weekly Supplement* of the *Survey of Current Business* for May 18, 1951, the figure for per cent of capacity operations in the iron and steel industry for April 1951 was 103 per cent.

electric to approximately zero. By 1940, Bessemer had fallen to 3,708,784 short tons, open hearth had risen to 61,744,889, and electric to 1,539,546 short tons. The shift from Bessemer to open hearth began just before World War I.[9] The point here is that there was Bessemer capacity, little used but available, and that in an emergency a good percentage of open hearth production could be turned over to idle Bessemer capacity, freeing open hearth for other purposes.

Dunn made the point, and emphasized it throughout his analysis, that capacity in the steel industry was set not only by ingot production: bottlenecks anywhere along the line could present a problem so serious as to invalidate whatever capacity estimates might be made.[10] For example, he cited inadequate mining machinery, marine or rail transportation, pig production facilities, scrap collection, or rolling mill capacity, all requiring attention in any meaningful analysis of iron and steel production. It was on this basis that he proceeded to make a detailed account of each step leading to actual ingot production. He followed this with estimates of capacity in the finishing and fabricating stages. The crux of these estimates was the figure 87,576,099 net tons per annum, which he calculated to be the maximum reliable industry capacity as of December 31, 1940. Once he had this figure he was able to determine, for example, what transportation facilities must of necessity be available to transport the required ore in order to produce the required pig iron which in turn had to be sufficient to meet the target figure of 87.6 million tons of steel.

The logic of the sequence can be seen in Table 10 in the

9. Dunn, "Report," p. 6. Recall that the Bessemer process involves air blown with great force up through a white hot molten mass of pig iron in a converter; the open hearth is a reverberatory furnace wherein hot gases are passed over the surface of molten metal composed of a certain percentage of scrap and a certain percentage of pig iron; the electric furnace melts the iron in the furnace by electric arcs.

10. De Chazeau made this point also: "Whenever one talks about the industry's ability to meet the demand, the demand may be for special products, while these capacity figures are general capacity figures. There may be bottlenecks in the ability to meet an increased demand for a special product." TNEC *Hearings,* 76th Congress, 3d Session, part 26, January 23–25, 1940, p. 10464.

tabulation of coke capacity. Here Dunn noted the blast furnace capacity needed for iron foundry and ferro-alloy products. Then, knowing that .88 of a ton of coke is required to make one ton of pig iron or ferro-alloy, he calculated the total coke required for these purposes. From the total coke capacity (less domestic heating, and so forth) he subtracted the foundry and alloy coke requirements. This indicated the coke available for steel-making pig iron. Then, referring to the 87.6 million ingot tons target, he indicated that 49.8 million tons of pig would

TABLE 10. *Tabulation of Coke Capacity as of December 31, 1940*[a]

	Tons per annum
Blast furnace capacity needed for iron foundries	8,000,000
Blast furnace capacity needed for ferroalloys	1,200,000
Total blast furnace capacity needed for these	9,200,000
Coke required to make one ton of pig iron or ferroalloy	.88
Total coke required for iron foundries and ferroalloys	8,096,000
Total coke capacity less domestic heating, commercial, etc.	47,395,812
Less required for iron foundries and ferroalloys (from above)	8,096,000
Coke available for pig iron used in making steel	32,299,812
Pig iron required for maximum reliable industry capacity of 87,576,099 tons steel	49,809,344
Coke required to make one ton of pig iron	.88
Coke capacity required for maximum reliable industry capacity of 87,576,099 tons steel	43,832,223
Coke capacity available for pig iron used in making steel	39,299,812
Deficit in coke capacity if maximum reliable industry capacity of 87,576,099 tons of steel is to be attained	4,532,411

[a]Dunn, "Report to the President."

be required to meet it. Multiplying this last figure by .88, he obtained the coke capacity required to meet the ingot goal. Finally, by subtraction of the coke capacity available for pig iron used in making steel from the coke capacity required, his figures yielded the deficit (in this case) in coke capacity, if the maximum reliable industry capacity of 87.6 million tons of steel were to be reached. This same approach was applied to determining whether there were a surplus or deficit in transportation facilities (he found a surplus of 1 per cent), blast furnace capacity (deficit of 3 per cent), and rolling mill capacity (surplus of average 15 per cent).

Calculation of scrap supply was ignored as too difficult to calculate, because of the many factors involved—that is, the price of scrap, transportation, and so forth. Dunn concluded that scrap would be available if the price were met. Here we have an example of potential error in this type of analysis, i.e., by-passing a relevant figure in a complex situation; for as we know now, scrap was a serious and definite bottleneck in war production. On the basis of his calculations, Dunn figured that of the needed 46,713,878 tons of scrap, over 26 million tons could be yielded as a by-product of the various processes in the mills themselves. This left approximately 20.5 million tons needed from scrap markets. He concluded that, as the market provided 18.2 million tons of scrap in 1937, the roughly 20 million tons would be easy to get in 1941. He said: "The industrial history of the country has shown that in any period of high production, increased scrap becomes available as a result of that high production."[11] It must be realized, however, that increased availability might be, and in the 1940s was, insufficient to meet the demands of the particular situation.

Dunn concluded his report by noting that "the readiness of alternative kinds of finishing capacity is essential to successful commercial operation. The sum of these alternative kinds, however, is invariable and necessarily the excess of the ingot capacity."[12] He stated that if the blast furnace and coke ca-

11. Dunn, "Report," p. 24.
12. Dunn, "Report," p. 25.

pacity deficits were taken care of, other facilities, including transportation, which led to the production of ingots and the capacity of those stages which led from the ingot output to the finished product, would be "in substantially harmonious economic relation to the ingot capacity itself." And it was on this premise that he based his conclusion, namely, that he was justified on these grounds in assuming that to determine ingot capacity plus capacity of certain steel foundries would give a correct and adequate measure of the capacity of the whole industry for the production of steel.

As an indication of the quality of Dunn's method of analysis, it is well to re-examine the basis upon which he judged the adequacy of capacity. Dunn spent, he stated, more than a month actively checking all the major steel companies with reference to their ingot producing capacity. He did this by asking the companies to report their capacity, as measured by "most economical production" and also their capacity as it might be "called into demand by the exigencies of national defense."

In the latter part of his report, Dunn turned to the reconciliation of his capacity estimate with estimates of total expected demand for steel. He referred to the de Chazeau Report,[13] and used it as his authority for "the estimate I have used of the amount of civilian consumption that may be expected to accompany increased national income." Dunn accepted de Chazeau's figures for anticipated civilian consumption, but converted them from "finished steel estimates" to "ingot estimates" on the basis of one ton of ingots to .72 of a ton of finished steel. The breakdown can be seen in Table 11.

Basing his estimates on a 61 million ton estimated civilian requirement for 1940 with an 80 billion dollar national income, Dunn established that a 10.1 million ton surplus could be expected for the year 1941. This can be seen in Table 12 and Figure 7; Table 12 includes his estimate for 1941, in which he assumed a 90 billion dollar national income and a corre-

13. Issued by the Office of Production Management, Washington, D.C., as *The Adequacy of Steel Ingot Capacity*, December 10, 1940.

sponding 70 million ton estimated civilian steel requirement. In the 1940 case, he assumed a reliable capacity of 87.6 million tons; in the 1941 case, he assumed capacity would have been expanded by 3.5 million tons. The actual figures for these years diverge significantly from Dunn's estimates. Reliable capacity was 83.6 million tons in 1940, but operation was at only 82.1 per cent of capacity. In 1941, reliable capacity was 87.3 million tons, but operation was 97.3 per cent of capacity. National income in 1940 was 77.6 billion dollars, and in 1941 was 96.9 billion dollars.

TABLE 11. *De Chazeau Estimate of Civilian Consumption*[a]

Civilian requirements by industry	*Fiscal year 1941*
Automotive	8.5
Construction	8.4
Railroad	5.2
Machinery	2.6
Container	3.4
Oil, gas, and water	2.9
Agriculture	2.5
All other	10.4
Total requirements (finished steel)	43.9
Ratio of conversion of finished steel to ingot assumed at .72	
Ingot steel equivalent of above total civilian requirements	61.0

[a]Dunn, "Report," p. 57.

Dunn pointed out that the liability to variation in his figures at different levels was attributable principally to the figures for civilian consumption, and he pointed out that 79 per cent of the total requirements would fall in this category. He said: "There is the variable involved in determining what proportion of the national income goes into steel under the changed industrial and economic conditions brought about by national de-

TABLE 12. *Final Determination of Surplus Steel Industry Capacity with Variations Resulting from Variations in National Income and Other Factors,*[a] *1940 and 1941*

As of December 31, 1940, national income 80 billions	*Millions of tons*
Direct defense requirements, including those of the Maritime Commission	3.1
Total export requirements, including those to Great Britain and Canada, in which is a considerable margin for possible underestimation	13.4
Civilian requirements, adopted from the de Chazeau report	61.0
Total of all requirements	77.5
Reliable capacity of steel industry found in Part I of [Dunn's] report	87.6
Estimated surplus of steel industry capacity as of December 31, 1940	10.1
As of December 31, 1941, national income 90 billions[b]	
Direct defense requirements, including those of the Maritime Commission	4.5
Total export requirements, including those to Great Britain and Canada, in which is a considerable margin for possible underestimation	14.5
Civilian requirements, adopted from the de Chazeau report	70.0
Total of all requirements	89.0
Reliable capacity of steel industry found in Part I of [Dunn's] report	91.1
Estimated surplus of steel industry capacity as of December 31, 1941	2.1

[a]Dunn, "Report."

[b]For the calendar year 1942 and assuming national income reaches 90 billion dollars.

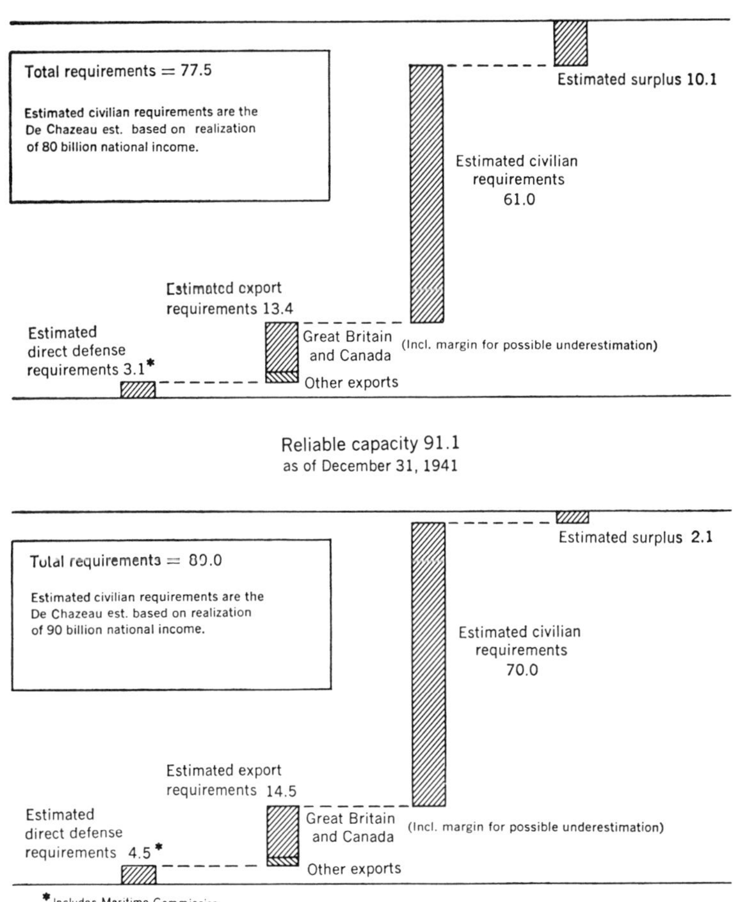

Figure 7. Reliable Industry Capacity and Estimated Requirements, 1940 and 1941 (millions of tons)

fense; and in addition there is the variable depending upon the amount of the national income itself."[14] The impact of restrictions on civilian consumption was reflected in Office of Production Management figures (Table 13) which show that the greatest effect on civilian consumption would be in the auto-

14. Dunn, "Report," p. 62.

TABLE 13. *Estimated Steel Requirements for Civilian Industries, September 30, 1941*[a] *(millions of new tons of ingot equivalents, calendar years)*

	1940 estimated consumption[b]	*1941 unrestricted requirements (national income $87 billions)*	*1942 unrestricted requirements (national income $100 billions)*	*1942 restricted requirements (national income $100 billions)*	*1942 per cent change from unrestricted to restricted requirements*[j]
Total	56.6	70.8	86.4	63.5	−27
Construction	10.8	12.9	15.4	9.8[c]	−36
Railroad	5.7	10.1	14.0	16.0[d]	−14
Automotive	11.7	12.9	15.1	5.3[e]	−65
Passenger				2.7	
Truck				2.6	
Machinery	2.6	3.5	4.5	4.5[f]	0
Agriculture	2.9	3.9	4.9	2.9[g]	−41
Container	5.0	5.4	6.0	5.4[h]	−10
Oil, gas, and water	3.9	4.9	5.8	5.8[f]	0
All other	13.9	17.2	20.7	13.8[i]	−33

[a]Bureau of Research and Statistics, Office of Production Management.
[b]1940 domestic consumption distributed, by industries, according to 1939 pattern.
[c]75% of 1941 requirements.
[d]1942 unrestricted estimate did not include expanded freight car program. *This* figure covers 222,000 cars.
[e]Based on 1.49 million passenger cars (65% reduction from 1941) and 820,000 trucks for civilian use.
[f]Unrestricted requirements. [g]Frozen at 1940 consumption. [h]1941 requirements.
[i]40 per cent of these requirements estimated to be consumer durable goods which will be curtailed by 50 per cent of 1941 requirements.
[j]Computed by National Association of Manufacturers.

motive industry, next in agriculture, and thirdly in construction. Railroads were the only units positively affected by such restrictions.

Dunn issued a second report in 1941, three months after the one discussed above. In the second he noted a considerable change in the various needs expressed by the different sectors of the economy. He found very little change in original capacity estimates. He pointed out that considerable differences existed between Office of Production Management civilian consumption figures and those of de Chazeau: the Office of Production Management's figures were derived by relating civilian consumption to increased national income, and the national income was anticipated to be 87 billion for 1941 and 100 billion for 1942. Dunn went on to say:

> I cannot place confidence in the validity of the relation between civilian consumption and national income from which these civilian consumption figures are deduced. In normal times it is possible that a relationship which has developed through a number of years is sufficiently stable to enable predictions for several years in the future, but I am convinced that in times like these, the relationship is unstable and the prediction correspondingly uncertain. The income and consumption figures undoubtedly rise and fall at the same time, but their predicted quantitative relation cannot be depended upon.[15]

He made the important point that although this type of estimate was valid with steel consumption rising faster than national income rises, it was important to remember that in such times as he was discussing and in many others, especially when there was full employment, there would be specific restrictions on consumer demand.

Dunn also asked a committee of the American Iron and Steel Institute to make an analysis of labor requirements for pro-

15. G. Dunn, "Second Report to the President of the United States on the Adequacy of the Steel Industry for National Defense," (Washington, Office of Production Management, Production Division, May 22, 1941).

ducing specified amounts of steel. Labor, of course, involved not only labor in the industry proper but in iron mines, transportation, blast furnaces, coke ovens, finishing mill operations, and also in industries consuming the additional steel that was produced, and in other secondary expansions that take place as a result of expansion in steel. Based on a consumption level of 101.9 million tons of steel in 1941 and 120.4 million tons in 1942, there would be a requirement, in 1942, for example, of 13,638,700 employees. This represented a 6,047,200 increase over the number employed for the same purpose in 1940. Dunn concluded that the labor market would have great difficulty in supplying this manpower; and, of course, with the advent of the war, which he did not consider, there would be an even greater problem. We will delay comment and evaluation of Dunn's approach until after a discussion of reports by Hauck and Bean, which provide some basis for contrast.

In September 1941, following the Dunn Reports, a schedule of plant expansion plans was submitted to the Office of Production Management by W. A. Hauck. Its general effect was to give impetus to a 10 million ton expansion of steel facilities. It created considerable controversy, with the industry generally opposed to what, at that time, seemed to be an inordinate amount of added capacity. The grounds on which the opposition pressed its case were set forth in a National Association of Manufacturers' paper on the iron and steel industry. Their claim that the recommendations of the Hauck Report were hard to understand rested on three points: (1) the record output of the steel industry, (2) the leading position of the United States in world production of steel, and (3) available data for over-all steel requirements did not, according to them, indicate a need for expansion on a large scale.[16] These arguments have, in part, been answered above; events proved the third to be wrong; and the expansion program was undertaken and exceeded in the decade following its issuance. But the Hauck Report was significant in its attempt to specify, by steel types, the proposed composition of the 10 million ton expansion pro-

16. National Association of Manufacturers, *Iron and Steel Industry*, pp. 5, 50.

gram. For example, Hauck specified additional ingot capacity for alloy steel of one million tons; for Bessemer steel, one million tons; for plates, one million tons; for West Coast capacity 1,865,000 tons; for balancing facilities 500,000 tons; for rolled armour, 100,000 tons—a total of 5,465,000 tons. In addition to this, he added projects under construction of 2,861,200 tons and an unapplied balance of 1,673,500 tons, yielding a grand total of ten million tons.[17] This approach was a move in the right direction; it pointed to rational expansion of the industry through the use of a detailed analysis of the product mix. Its advantages are considerable, compared to more aggregative estimates of steel plant capacity.

The war period was one of high levels of capacity utilization and considerable expansion of capacity. In 1940, capacity stood at 81.6 million net tons; by 1945, it had reached 95.5 million tons. In 1946, considerable equipment that had been kept in use during the emergency was scrapped, with a resultant reduction of capacity to 91.9 million tons.

The bases for more recent estimates of capacity needs, including those made by the industry to substantiate its position that adequacy was achieved by the end of World War II, are to be found in the hearings of the Senate Small Business Committee. Among those testifying from the industry's vantage point was Wilfred Sykes, then president of the Inland Steel Company. He appeared before the sub-committee in the summer of 1947 and stated: "The current steel capacity, even without improved facilities and processes, would be adequate for all potential demand."[18] Sykes spoke in per capita terms and mentioned that per capita consumption of steel had ranged between 236 pounds in 1932 and 978 pounds in 1929. Assuming 978 pounds per capita consumption with an additional 10 per cent for export, he stated that the maximum steel consumption

17. W. A. Hauck, *Steel Expansion Program Submitted to the Office of Production Management* (Washington, September 24, 1941). See also W. A. Hauck, "Status of the Steel Expansion Program," (War Production Board, Washington, June 30, 1942).

18. U.S. Congress, Senate, *Interim Report of the Special Committee to Study Problems of American Small Business,* 80th Congress, 2d Session (Washington, Government Printing Office, January 16, 1948), p. 27.

in 1950 would be 76,373,000 tons, in 1955 it would be 78,-464,000 tons, and in 1975 it would be approximately 90 million tons. He then added that all of this was below the present theoretical capacity.[19] His estimates are indicated in Figure 8.

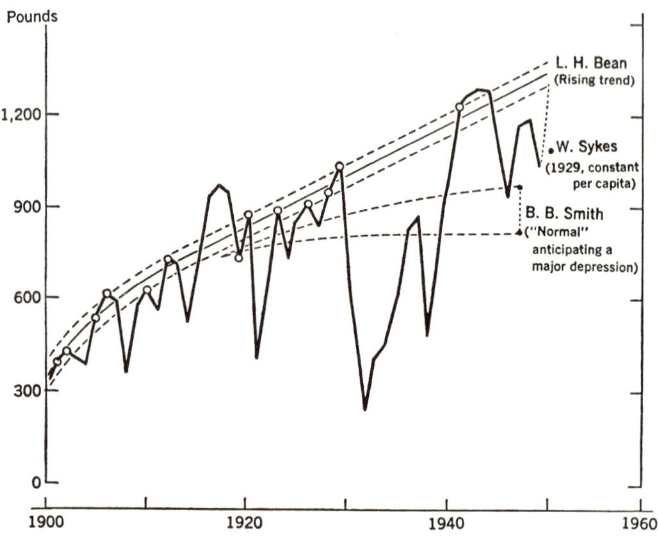

Figure 8. Three Estimates of Steel Production (ingots and castings) per Capita Required for Full Employment, United States (Source: U. S. Department of Agriculture)

In his testimony, Louis H. Bean presented conclusions regarding the amount of steel production required to maintain full employment and a high level of industrial prosperity. Bean also used per capita analysis of steel needs. He stated that by 1900 the industry had expanded to the point where one-half ton per person was employed throughout the United States. By 1912–13, production was equal to nearly one ton per job, by 1929 it rose to 1.35 and by 1941 to 1.7 tons per job. He claimed that if one takes this trend in steel production per job opportunity in prosperity years at 1.8 tons for the years 1947

19. Sykes used population figures of the U.S. Bureau of Census, which estimated population at 143,896,000 in 1950.

to 1950, a total employment of 58 million would require over 100 million tons of steel, compared with the annual rate of output of 85 million tons during the first part of 1947.[20]

Bean's recommendations included an increase of capacity to between 100 and 110 million tons in order to bring about full employment prosperity through 1947 to 1950. He projected his increased per capita demand for steel on the high points of production accomplished in prosperity years. This can be seen in Figure 8. Bean's trend figure for 1950 would be 1,350 pounds per capita, which, times an estimated population of 152 million, yields a necessary steel output of 100 million tons. He claimed, however, that such an estimate would press the industry very closely, and only if there were another 10 per cent to provide for shutdowns would conditions be satisfactory.

Ewan Clague, of the Bureau of Labor Statistics, also appeared before the Committee. His testimony was based for the most part on a study the Bureau had made of estimated levels of economic activity for a full employment economy in 1950. The method employed was essentially the Leontief input-output technique, using flow coefficients and basing everything on full usage of the expected labor force for 1950.[21] The system was worked out for two models, a consumption model and an investment model. The former yielded an estimate for steel production of 98 million tons, and the latter one of 120 million tons. These estimates have subsequently proved to be as close to what actually occurred as any that were made at the time. It is important to note two assumptions that underlie the type of analysis undertaken by the Bureau of Labor Statistics: (1) that a high level of economic activity with high efficiency will persist; and (2) that the behavior habits and working hours of the consumer will remain constant.

Additional attention was paid to questions of adequate capacity in the Judiciary Committee hearings on monopoly power in 1950. Four witnesses, Earle C. Smith, chief metal-

20. U.S. Congress, Senate, Small Business Committee, *Interim Report,* p. 28.

21. Exposition of this Bureau of Labor Statistics' study appeared in the February and March 1947 issues (vol. 64) of the *Monthly Labor Review.*

lurgist of Republic Steel, James Boyd of the U.S. Bureau of Mines, Louis H. Bean, and Kenneth Hunter of the Economics Department of American University, estimated future demand for steel (Table 14).

TABLE 14. *Estimates of Steel Ingot Capacity in the United States, 1950–70*[a] *(millions of net tons)*

Year	*Smith*	*Boyd*	*Bean*	*Hunter*
1950	99.4	99.4	100	
1951	100.0	100.0	102	
1952	101.0	101.0	104	
1953		101.0	106	
1954		101.0	108	
1955	102.0	102.0	110	110[b]
1956	103.0	103.0	112	
1957	104.0	104.0	114	
1958	106.0	106.0	116	
1959	108.0	108.0	118	
1960	110.0	110.0	135	
1965	117.0	117.0		
1970	125.0	125.0		

[a]U.S. Congress, House of Representatives, *The Iron and Steel Industry,* Report of the Subcommittee on Study of Monopoly Power of the Committee on the Judiciary, 81st Congress, 2d Session, December 19, 1950 (Washington, Government Printing Office), p. 68.

[b]"In the near future;" no specific date given. *Hearings before the Subcommittee on Study of Monopoly Power,* p. 776.

These estimates and the methods employed, however, do not close the demand measurement discussion. Another attempt to analyze demand for steel is to be found in the proposal and subsequent study of the feasibility of erecting an integrated steel works in the New England region. It is of interest not only because it represents an example of an attempt to assess cost of construction but also because it introduces the regional market aspect of the problem.

Acutely aware of its position as the "mature" region in the American economy, conscious of the migration of many textile establishments to the South, facing a loss of markets to competitors from the Middle and Far West, leaders in New England felt strongly about the desirability of having steel-producing facilities in their region. The growth of metal fabricating industries in the area and the post-World War II pattern of generous government loans to finance the building of plants contributed to raising interest in this particular project. In addition to study, by government and business leaders from the various states, of the feasibility of a New England steel mill, organized inquiry was pursued by the New England Council and the Federal Reserve Bank of Boston.

Among the arguments put forward in support of a New England mill was the assertion that New England metal-working industries were at a competitive disadvantage because they had to pay higher freight rates than manufacturers using steel in other parts of the country. This was made more severe by the change from basing point pricing to f.o.b. pricing (under which the steel manufacturer absorbed some of the freight costs in New England).[22] Another development which was to the disadvantage of New England steel consumers was the expansion of new mills on the Middle Atlantic coast and in Utah, Montana, and California. The contention was that all these mills were supplying steel to consumers near to their location at prices substantially lower than those which New England steel users paid. It was claimed that in periods of steel shortage, New England producers were at a disadvantage, that they were "allocated" inadequate amounts of steel for their needs. The effect of this actual or potential steel shortage in New England was said to carry over in discouraging users of steel from locating production facilities in New England and also in encouraging presently located New England fabricators to move nearer to ready sources of supply, hence away from New England.

The claim that a New England mill would be profitable was

22. These arguments were presented, for example, in C. Bowles, "The Prospects for a Connecticut Steel Mill: A Report to the People of Connecticut," Hartford, February 1950.

based on the fact that adequate construction sites and a labor supply were available, that an adequate supply of raw materials at costs competitive with those available to producers in other areas existed, and that the market which a New England mill could serve would be sufficiently large to make the venture profitable. In Connecticut, a Steel Advisory Committee of leading businessmen was appointed by the Governor in October 1949. Some five weeks later, the Committee reported back to the Governor that they believed that they should make inquiries of major steel companies as to whether they would have any interest "in locating in Connecticut and aiding in the financing and operating of such a mill." The Governor's Report indicated that a 1.25 million ton integrated mill would cost at least $200 million; in addition, $50 million would be needed as working capital. A temporary loan from the Reconstruction Finance Corporation was under discussion. There was also discussion of building a mill to the specifications of an operating company with government funds and then renting the facilities to the operating company, allowing them to purchase it over a period of years.

The Governor's Report discussed the relative merits of inducing already existing large companies to come into New England and operate a plant, or starting an entirely new corporation, with New England financing. It was acknowledged that the great advantage of bringing in an operating company would be to take advantage of the managerial skills and experience of such an organization. It was also acknowledged that Bethlehem would not be a logical company to approach; with their Sparrows Point mill, another Atlantic Coast outlet would be of little interest to them. At the time, U.S. Steel's plans for the Fairless Works were known, but no action had been taken; the report indicated that the Fairless plant would give U.S. Steel a preferred position in New England as long as no other New England mill existed, so U.S. Steel was grouped with Bethlehem as unlikely to show any interest in the New England proposition. The report went on to say:

> The attitude of the existing companies is by no means fully crystallized. But as we anticipated, there was a very understandable reluctance to upset the present distribution and production pattern. Since these producers are already selling steel in the New England market from their present locations at a generous profit, there was no real incentive for them to build a mill in this area.

Though the report stated that steel companies did not claim a New England mill would be unprofitable, they indicated that they had other uses for their investment funds, in terms of national plans which would be more profitable to them. As a result, the report concluded that if the mill were to be built in New England, it would have to be built with New England capital and under New England management. It suggested that a complete and detailed economic and engineering survey be conducted to establish additional facts on costs and site planning for a possible New England mill. About $60,000 was then appropriated for the purpose of conducting such a survey.

In January 1951, the New England Steel Development Corporation of Hartford, Connecticut, as the promoters of the project were known, was reported to be within thirty days of completing arrangements for the construction of a million ton ingot facility.[23] The government had certified some portion of the $250 million investment for five-year accelerated amortization; this had the effect of insuring bank and insurance company loans as well as equity financing for part of the construction fund. There would also be financial aid from the Reconstruction Finance Corporation. The mill was to be located on a 1,600 acre waterfront site west of New London, Connecticut. The facility was to be primarily a producer of flat rolled products, which would put it in direct

23. "Steel Mill Likely for New England: Federal Grant of Tax Benefits Eases Way for Plant with Million-Ton Capacity," *The New York Times* (January 14, 1951) p. F1.

competition with the already existing Bethlehem facility.[24]

On March 9, 1951, an item appeared in *The New York Times* noting that the engineering firm of Coverdale and Colpitts had issued a report on March first, which was being recalled at the request of Governor Lodge of Connecticut and the President of the Connecticut Steel Advisory Committee and the New England Steel Development Corporation. The ostensible reason for recall was "additional information" which had come to the attention of the engineering firm. The news item continued: "According to political sources which have read the report, it is highly unfavorable to the proposed steel mill, declaring there would be no New York market, iron ore would be hard to get, the proposed mill would be unprofitable for at least two years, and coal would be difficult to secure."[25] It might be useful to examine the report as it was finally issued on April 17, 1951.[26]

The March first report was divided into two parts. The first was a "Report on Prospective Markets for Steel Mill Products." It included a review of steel-making capacity in the United States and then proceeded to discuss the possible markets for a New England plant. A review was made of consumption possibilities in New England proper, in metropolitan New York, and in New Jersey. The review was based not only on statistical study but also on field work. It consisted of a close study of the Bridgeport metropolitan area, considered to be a "representative metal-working center," and of interviews in other market areas. "During the course of these field surveys we made a total

24. At about the same time, an application for a $100 million loan to aid in the construction of a steel plant near Clinton, Iowa was reported. Plans were for production of one million tons of finished steel and 400,000 tons of pig iron annually. There was also a report of a government loan to a Kentucky firm of $8,356,000 to construct a plant with an annual ingot capacity of 189,000 tons. "$100,000,000 Sought for Steel Plant: Iowa Industrialist and Government Officials to Confer on Plea for Loan," *The New York Times* (February 8, 1951) p. 41.

25. "Engineers Recall Report on New England Steel Mill," *The New York Times* (March 9, 1951).

26. "Report on Proposed New England Steel Mill for New England Steel Development Corporation," Coverdale and Colpitts, Consulting Engineers (New York, April 17, 1951).

of more than 500 calls." Beyond this, the report included a discussion of the types of steel plant facilities that would seem appropriate. The engineers recommended facilities to turn out products embracing a large part of the market, but of necessity their recommendations precluded satisfying certain parts of the market, namely: structural shapes, for which the largest consumers are in the New York-New Jersey area (the New England producer would be at a competitive disadvantage); rod and wire, 75 per cent of which was already produced in New England and hence the New England mill would then be competing with other New England producers; bars, for some of which electric furnaces are not best suited. The remaining types (carbon and reinforcing bars) were effectively distributed by the already existing warehouse facilities in the area. Plates, which would require special finishing equipment, were precluded; so were specialty sheet and strip, such as electrical enameling, etc., which cannot be produced on standard rolling mills with standard processing equipment. What was recommended was a "mill producing rolled flat products," because (at the time this study was made) the shortage of these products was marked, and the portion of flat rolled products in total steel rolled products was increasing. The tin plate business was important in the area, and a stable item in the steel product picture; further, the "profit margin in these products is generally higher than in shaped products." The recommendation, then, was that the New England mill produce hot rolled sheets and strip, cold rolled sheets, tin mill products, and butt weld pipe.

In examining this report, there are two factors which should be noted. First, these estimates of markets were made in 1950, when steel was in short supply, and when the picture of final demand for steel products might have been distorted due to defense expenditures. Also, as the report stated, the estimates of markets and suppliers in these markets were made before the effects of the output of the Fairless Works could be felt in the area; it would have been quite heroic for the New England Development Corporation to have taken these estimates at face value, without further evidence of the effects of expansion at Sparrows Point, the Morrisville works, and the possibility of a National Steel plant at Camden on their potential market. The engineers did make a point which is particularly appropriate in the context of steel consumer behavior:

> It must be realized that many consumers, particularly the larger ones, insist on more than one source of supply. Other consumers have historical sources of supply, and their value added in manufacture is such that the extra cost of their steel from a more distant producer is not of prime importance. This is particularly true in the New England market. There are many other factors, such as the desire to purchase steel from special quality requirements, etc., which narrow down the potential participation of a New England mill.[27]

Another consideration in appraising the approach used by the authors of the report is their treatment of specialty needs in the consumer market. Insofar as the roster of consumer items consists of many specialized items, many of which were specifically excluded in the Coverdale and Colpitts study, the "ability" of a New England mill to satisfy New England demand for steel products is a generalization that should be highly qualified. Studies of costs of materials (including transport costs) which, in the aggregate, suggested the advantages which a mill at New London would have in satisfying that market had to be tempered by the acknowledgment that it undoubtedly would not be producing the full range of commodities consumed in Bridgeport, and hence, the dollar cost advantage that might be expected in the assemblage of materials and shipment of a finished product to the market could be evaluated as only part of the story. Insofar as the New London mill would be limited, for example, to the production of the line of products recom-

27. Ibid., p. 17.

mended by the engineering firm, the impressive figures on cost advantages which might be enjoyed by the New England mill still might not justify its construction.[28]

The second half of the Coverdale and Colpitts report consisted of a survey of specific iron ore, coal, limestone, and steel scrap sources, a description of steel mill facilities which would be recommended, including a raw materials dock, coke ovens, blast furnace, open hearth shop, soaking pits, blooming mill, hot strip mill, cold rolled sheet and tinplate mill, pipe mill, electric power, water supply, and other facilities. This was followed by an estimate of construction costs and then a reconciliation of production costs and earning capacity. In their summary, the engineers noted that they based their calculations of cost and profitability on the assumption that the proposed mill would be an independent operation, securing its own raw materials with its own administrative, operating, and sales forces. They went on to say:

> We wish to point out that the prospects of such a mill are considerably less favorable, at least in the first few years of its existence, than if it were built and operated as part of an existing large steel company. The problem of design, construction, and starting-up would be less. Acquisition of raw materials would be greatly aided. Probably the most important favorable factor would be the more rapid development of the market and thereafter the maintenance of a greater market than would be available to an independent mill. Economies in management and other general expense should be realized. It is not possible to estimate what the total advantage would be, but

28. For an analysis of cost factors, transportation costs of input, comparative rail and motor carrier rates, scrap prices and a breakdown of New England consumption of steel, production of steel and consumption of non-New England steel and conclusions as to the feasibility of a New England works, see Walter Isard and John H. Cumberland, "New England as a Possible Location for an Integrated Iron and Steel Works," *Economic Geography, 26* (1950), 245–59.

> the difference might be as much as twenty-five per cent in operating profit.[29]

As was indicated above, the main body of this report was submitted on March first to the New England Steel Development Corporation. A letter accompanying the resubmission on April 17th made the following point:

> An arrangement had been concluded in November of 1950 that would result in the engineers' report, when ready, being submitted to the Bethlehem Steel Company at the same time as to the New England Steel Development Corporation. This arose out of an offer on the part of the Development Corporation to any company interested in exploring the possibility of a New England mill and cooperating with those making the study to have an option on information derived in the final study and an option to take over the project in the event that the company was interested in doing so. Bethlehem went along with this and cooperated with the consulting engineers. Bethlehem agreed to make a proposal for consideration by the Development Corporation within a reasonable length of time after the submission of our reports or to withdraw from the project promptly, in which event the Development Corporation would be free to use these reports in negotiation with other companies or in carrying out the project itself.[30]

The Development Corporation met with representatives of Bethlehem on March fifth, and Bethlehem announced that it was not interested in the project and waived its option. Bethlehem's spokesman did not say that the company might not be interested in reopening negotiations at some future date.

In examining the consulting engineers' report, the directors of the Development Corporation on March eighth questioned

29. Coverdale and Colpitts, "Report."
30. Coverdale and Colpitts, Letter of Transmittal, April 17, 1951, p. 2.

some of the findings of the study, notably with regard to figures on raw material costs and the New England market, and also raised the question of whether the engineers should not have devoted some additional attention to the effects of the Corporation having obtained a certificate of necessity, and whether this would be of interest to existing steel companies in operating the proposed New England facility.

The engineers explained the withholding of the report after its preliminary release as an attempt on their part to examine some of the questions raised by the Development Corporation. In their letter, the engineers noted: "The one main obstacle to the construction of the mill at present is the capital cost from the ground up. This same obstacle stands in the way of new integrated steel mill construction anywhere, particularly for a new and independent company."[31] They went on to note that in comparing the operations of the industry in the average versus that of the independent New England company, though the operating profit ratio might be the same, "the heavy burden which a new mill would have to carry in depreciation charges, as compared with the industry or any particular company that operated a plant built mostly at pre-war costs," would be markedly different, i.e., net income available for interest and taxes for the steel industry together average 12.5, and for the independent New England company 5.3, as per cent of sales. The engineers pointed out that although the certificate of necessity which the New England Steel Development Corporation had obtained would be of only "limited value in reducing the capital cost of the project, because such a new company could not be expected to earn profits in the first five years to realize sufficient tax savings through accelerated depreciation to make it a financially attractive project," the value of such a certificate to an already existing steel company (which could file a consolidated tax return) would be quite different, and might be very attractive.

On April 20th, a letter was sent to Governor Lodge by the chairman of the Connecticut Steel Advisory Committee, in

31. Ibid., p. 4.

which he noted that his committee felt: "(a) that a steel mill independently operated would be unjustified economically and should not be encouraged and (b) that if an existing well-qualified company should express an interest in building and operating a steel mill in Connecticut that our state should encourage such a venture."[32] It therefore appears that the sequence of events was such that the emergence of a change in locational pulls in the industry, shifting from old favored sites (basically coal-producing sites) to market sites for the product, combined with the discovery of ore resources in Venezuela and Labrador, made an East Coast port location highly desirable, and stimulated activity in the New England area for construction of a facility in that area. This undoubtedly stemmed from the complicated changes that were taking place in the New England economy as well as the regional chauvinism which so often stimulates bursts of "conspicuous" economic activity, with political overtones. The coincidence of these technological factors with the financial possibilities suggested by accelerated amortization and federal financing made the prospects for the success of this venture quite bright. Two factors worked in the other direction, however; there was the existing market structure of the industry, but more to the point, the plans of existing large producers either to expand or to begin from scratch facilities which would be operating adjacent to this market area, and which would be realizing similar benefits from the new sources of ore supply. Finally, though supported by government finance arrangements and considerable regional desire to see the project under way, profit expectations for an independently sponsored venture were so low that the project was postponed.

The certificate of necessity originally issued for the mill expired on August 11th, 1952, and it was not until October 1953 that further efforts were made by the New England Development Corporation to obtain a new certificate of necessity. However, it is interesting to note that this time, rather than a project

32. Letter from Clifford S. Strike, Chairman, Connecticut Steel Advisory Committee, to Honorable John Davis Lodge, Governor of the State of Connecticut (Hartford, April 20, 1951).

whose estimated cost would be in the vicinity of $228 million, the new certificate enabled the owners to write off 65 per cent of a $26.4 million project. "The proposed plant was described as consisting of two large electric furnaces for steel ingot production, and a rolling mill for converting the ingots into finished steel products."[33] The project also stood pending for some time, and to the date of present writing, due in part at least to some of the factors indicated in the foregoing, New England is still without its longed-for fully integrated works.

This concludes a survey of some of the previous attempts to analyze the demand for steel. Mention might well have been made of other attempts, such as those of the National Industrial Conference Board[34] and the Paley Report,[35] but the studies which we have examined serve to indicate the limitations of

33. "Tax Write-Off Set for Steel Plant: O.D.M. Grants Benefits for New $26,400,000 Mill Projected in New England Area," *The New York Times* (October 31, 1953) p. 22.

34. In 1951, the National Industrial Conference Board released a study of steel demand in which they raised the question, "how will production of close to 90 million tons (of finished steel) square with the steel needs of 1953?" After reconciling alternative amounts to be allocated for military steel requirements, they indicate: "there is even a chance . . . that actual steel demand might fall below available supply. This might happen if sufficient manpower were lacking to work the steel. During the last war the number of man-hours required to fabricate a ton rose sharply, owing to the specialized requirement of a war time economy. If this were to happen in the next couple of years, the manpower problem could become critical." Cf. "Ample Steel Seen to Fill All Needs," *The New York Times* (April 16, 1951) p. 32.

35. Under a discussion of "key" commodities, the Paley Report briefly reviews the history of expansion in the iron and steel industry over the period 1925 through 1952; it then recalls the debate in the post-World War II period on future requirements for capacity. It notes: "The outlook has now been altered by the course of events. Today, the industry is confident that capacity will approach 120 million tons in 1953. And this figure does not include the mill that has been proposed for New England or a number of projects that are under discussion elsewhere." It then goes on to discuss the costs of expanding capacity, and indicates that whereas before the war an integrated plant could be built for less than $100 per ton of finished steel, currently the cost would be close to $350. It observes: "With respect to demand, the industry has recently raised its sights." *The Outlook for Key Commodities, Resources for Freedom, A Report to the President by the President's Materials Policy Commission, 2* (Washington, Government Printing Office, 1952), 17. Cf. also Clarence E. Sims, "The Technology of Iron and Steel," in *The Promise of Technology, Resources for Freedom, A Report to the President by the President's Materials Policy Commission, 4* (Washington, Government Printing Office, 1952), 31–44.

statistical techniques and the complexity of the variables which make quantitative analysis of demand for the purposes of capital expansion so difficult. The state of affairs as to the reliability and usability of these techniques was aptly conveyed in a summary comment of the Senate Small Business Committee's examination of steel supply and distribution problems as early as February 1949. The most conspicuously repeated sentiment in their Final Report was that they lacked accurate measures of what conditions actually existed in the industry, and then they added this specific judgment:

> There is no reliable measure of present demand for the important basic materials of industry as related to the capacity to produce them. Likewise, we do not know what volume of these commodities is required to support the high levels of production and employment established as national policy under the Employment Act of 1946.[36]

INVESTMENT DEMAND, PROFITS, AND THE ACCELERATOR

It is necessary at this point to return to a question of emphasis raised in Chapter 2. Implicit in much testimony on the demand appraisal issue has been the assumption that motivation to increase capital expenditures will, for the most part, stem from a rise in output in the economy—the acceleration principle. It has been said that demand for steel is not direct, that it depends rather on activity in automobiles, machine tools, construction, electrical equipment and so forth. At the same time, however, we must note that when this picture is cast in a broader general equilibrium framework, the distinction between derived and direct demand loses much of its meaning.

36. U.S. Congress, Senate, *Final Report of the Special Committee to Study Problems of American Small Business,* 80th Congress, 2d Session (Washington, Government Printing Office), p. 26. In its *Report,* this Committee attempted to set up a framework for estimating iron and steel capacity. In many respects it resembles Dunn's approach.

In essence, all demand is derived; it arises from movement in one or another sector of the economy.

The question of perspective can be said to have a very important part in the various interpretations of the economy's needs for steel capacity. On the one hand, demand for steel can be seen as the obvious reaction of the consumers of steel to a change in the demand pattern in their own industry, or a specific change in the outlook of the decision-making entrepreneur. On the other hand, demand for steel can be seen as the necessary concomitant to change in general production levels or to changes in growth rates in the economy at large. Recent literature concerning demand conditions does contain examples of attempts to join these two divergent perspectives into a consistent and operative appraisal of the total demand picture.[37]

In discussing the accelerator, the question of excess capacity is raised once again. In Chapter 3, where we used a non-linear model to suggest the bottleneck potentialities of the industry, we did not include the possibility of less than full capacity utilization in the discussion because at that point we were concerned with the upswing of the cycle and the precipitation of a downward turning point. However, it is necessary, when

37. Among these are articles by Boschan, Clark, Klein and Christ: see P. Boschan, "Productive Capacity"; C. Clark, "A System of Equations Explaining the United States Trade Cycle, 1921 to 1941," *Econometrica, 17* (1949), 93; L. R. Klein, *Economic Fluctuations in the United States, 1921–1941* (New York, John Wiley, 1950); C. Christ, "A Test of an Econometric Model for the United States, 1921–1947," *Conference on Business Cycles* (New York, National Bureau of Economic Research, 1951). An earlier attempt at building an integrated system is found in J. Tinbergen, *Statistical Testing of Business Cycle Theories, 1, A Method and its Application to Investment Activity* (Geneva, League of Nations Intelligence Service, 1939), and *Statistical Testing of Business Cycle Theories, 2, Business Cycles in the United States, 1919–1932* (Geneva, League of Nations Economic Intelligence Service, 1939). A more recent study touching on many of the same questions as those Tinbergen examined is that of Meyer and Kuh. Their purpose was to study possible explanations of investment incentive through the use of cross-section data. They marked out three areas for testing: profit maximization, including modifications of "classical" marginal analysis to include considerations of uncertainty and expectations, the accelerator relation, and the drive to retain a market share. Some comments on their work appear later in this section. See J. R. Meyer and E. Kuh, *The Investment Decision: An Empirical Study* (Cambridge, Mass., Harvard University Press, 1957), p. 4.

viewing the overall cyclical picture and relating investment motivation to the total cycle, to include some additional remarks concerning capacity utilization and its relationship to the accelerator factor.[38]

Even before one gets to capacity utilization figures and questions of the significance of such ratios, the problem of measurement of capacity itself poses problems. Just what does a given ingot capacity figure represent? Reference to "rate of capacity" was made in connection with the Dunn figures at the beginning of this chapter, but more than this should be added. Such questions as the classification of both labor (in the sense of a guaranteed annual wage) and the capital equipment as fixed or variable cost items, the time intervals associated with the particular capacity figure, whether particular inputs to the industry are classified in such a manner that they must be considered "idle" when not being fed into the industry—all deserve attention.[39] Whether the capacity figures represent what is accepted by the industry as standby capacity and therefore not typically in use but ready and available, or whether quantities of obsolete facilities which, with some effort, could be reactivated but which typically are not regarded even as standby are included, or whether conversion of facilities devoted to other production should for some reason (particularly military emergency) be considered part of the capacity potential of the industry or not —all are complications. Once ground rules are agreed upon, they do not prevent specific measures of capacity being pinned down. They do, however, emphasize the necessity for indicating the capacity concept which lies behind a given set of figures. There is, for example, the possibility that a capacity figure represents the use of physical facilities to a maximum, yet without going to the limit which brings on inordinate deterioration; an alternative is a capacity figure which represents "optimal"

38. The relevance of consideration of all phases of the cycle will be illustrated in reference to investment behavior of the Inland Steel Company in Chapter 5.

39. Cf. E. Zabel, *Concepts and Measurement of Productive Capacity* (Princeton University, Technical Report issued under Office of Naval Research Contract, November 1955).

levels of output in an efficiency sense, i.e., utilization which brings production to the low point of the average cost curve. The former capacity concept would involve "over-driving" the system and hence implies "distortion" or increasing costs. From the economist's point of view, this would not be the useful capacity concept, but under certain conditions, e.g., war emergency, the costs incurred in over-drive might be justified in terms of deadlines for certain quantities of output.

In the case of the steel industry, not only is the physical dimension of the blast furnace or converter relevant, but the quality of ores, coke, and fluxing material as well; also the possibility of varying the ratio of scrap to pig iron and hence varying the output within the dimensions of a given open hearth furnace is relevant, and this argument can be carried further in terms of the "efficiency" of the labor force and to some (perhaps unreal) extent to the administrative efficiency of the larger organization itself. The point, of course, is that the physical entity is not the only consideration in the presentation of capacity figures for a plant; rather it is the complex interrelationship of the several variables and their use in a given situation that determines a usable capacity figure.[40] One complication that arises when planning new capacity is that productive potential of new furnaces is anticipated on the basis of historical experience with furnaces of roughly similar types and with corresponding adjustments for anticipated technological improvements. However, it is not until actual operation of the facility that the specific capacity figure in question can be derived; there is some evidence that, as a result of the estimation procedure, anticipated output from new facilities tends to fall short of what actually is realized, and hence capacity expansion figures ex ante tend to the conservative side. Existing facilities also receive increases in their rated capacity figure as a result of technological improvements. These involve either modifications of existing equipment or the modification of handling of input materials in some manner which results in a net gain

40. Note the calculation of several blast furnace capacity figures in Table 9, p. 95. Cf. also Zabel, pp. 43–45.

of output in the process. One way such gains are realized is through improved utilization of fuels and heating facilities that cut heat times with the same "volume" of input materials.

Chenery is among those who attempt to introduce capacity considerations in discussion of the acceleration principle.[41] He concentrates on the relationship between economies of scale and motivation to expand capacity, and makes the point that when firms are in a position to realize economies of scale because of industry cost conditions, they will tend to build in anticipation of increases in demand, with a resultant tendency for less than full capacity utilization in such industries. He suggests that the market structure of a given industry may be the key to whether the straight accelerator or one which includes allowance for capacity utilization is the appropriate variable in an analysis of investment behavior in particular cases. His findings lead him to the conclusion that in more highly concentrated industries and industries with high capital output ratios the capacity utilization logic holds, whereas in more monopolistically competitive industries or nearly competitive industries with lower capital intensity, the simple acceleration principle is appropriate.[42]

The main distinction, then, between the simple acceleration principle and the type of modification Chenery employed rests on "the fact that capital is not and cannot be contracted as rapidly as it expands. In each industry there is a technological limit to the rate at which capital is retired which depends on its durability, rate of obsolescence, and so forth."[43] He notes that

41. H. B. Chenery, "Overcapacity and the Acceleration Principle," *Econometrica, 1* (1952), 1–26. Cf. also A. D. Knox, "The Acceleration Principle and the Theory of Investment," *Economica,* N.S. *19* (1952), 279–80. Knox distinguishes between the accelerator theory being able to explain the timing as contrasted with the volume of investment; he gives a negative judgment on its abilities with timing, but acknowledges that the principle, as applied to explanations of the volume of investment activity, seems valid.

42. Meyer and Kuh indicate that the problem of spurious correlation may have led Chenery to believe the simple accelerator is more appropriate in the case of the petroleum industry where their findings lead them to feel otherwise. See E. Kuh and J. R. Meyer, "Correlation and Regression Estimates When the Data are Ratios," *Econometrica, 23* (1955), 400–16.

43. Chenery, "Overcapacity," pp. 12–13.

when the accelerator is applied over the full cycle it forecasts "too rapid a decrease of capacity in the downswing and too rapid an increase in recovery." In averaging over the cycle, the accelerator coefficient typically derived does not represent accurately the various phases of the cyclical picture. He proceeds to allow for this in his capacity approach.[44] In Chenery's analysis there is a modification of the acceleration principle based on an optimum relationship between the output of a firm and the capacity to produce that output. It is related functionally not only to such considerations as the planning period, the rate of increase in demand, and the discount rate, but also to economies of scale. "Once overcapacity exists it is taken into account in further investment and hence should be embodied in whatever equation is used to predict fluctuations in investment."[45]

These remarks on measurement of capacity and the possibility of modifications of the simple acceleration principle lead us back to some references to the empirical study of Meyer and Kuh.[46] Meyer and Kuh note that in cases where there is evidence suggesting not only cyclical overcapacity, but conditions of secular excess capacity—a condition which we have seen in the steel industry for some periods—"the level of output and the firm's capital stock become the relevant variables, rather than the change in output alone."[47]

Meyer and Kuh also raise the question of availability of financing as a major determinant in investment behavior. The apparent preference (or preoccupation) of large-scale enterprise for financing expansion out of retained earnings has been a prominent feature of many empirical studies on investment

44. Chenery ("Overcapacity," p. 13 and n. 20) calls attention to the fact that Kuznets "pointed out the inevitability of overcapacity in industries where the ratio of capital to output is large, and the correlation between capital intensity and the expected overcapacity." He goes on to note: "It is for these industries, where induced investment may be very large, that a modification of the simple accelerator is more needed." He cites S. Kuznets, "The Relation between Capital Goods and Finished Products in the Business Cycle," *Economic Essays in Honor of Wesley Clair Mitchell* (New York, Columbia University Press, 1935), pp. 209–67.

45. Chenery, "Overcapacity," p. 26.

46. *The Investment Decision: An Empirical Study.*

47. Ibid., p. 15.

motivation. In these studies, management statements concerning the liquidity position of the firm also indicate the importance of this factor.[48] Meyer and Kuh attribute these attitudes to: "(1) The disadvantages that arise when a firm extends its external debt position; (2) historical events and institutional adjustments which have made outside funds difficult and expensive to obtain; and (3) the hierarchical structure and motivation of corporate management which makes outside financing asymmetrically risky for the established or in-group."[49] They note, for example, that because of the fixed interest aspect of indebtedness it is "likely to involve restrictions on managerial flexibility and prerogatives," and that stocks are an "expensive method of raising money." In addition they note that the "trade position" of the firm is of great importance to the businessman. "He is driven by the desire to keep pace with his rivals, and investment is undertaken when and if needed to keep one's standing in the industrial hierarchy."[50]

As regards the liquidity aspect of investment motivation, Meyer and Kuh find from their study "(1) that the grant of accelerated amortization privileges has an influence on investment decisions even in non-inflationary times—if not more so; and (2) that the liquidity aspect of depreciation expense seems to be the variable's dominant characteristic."[51] They emphasize the point that it is the liquidity aspect rather than the relative durability of the capital equipment involved that affects the entrepreneurial decision to take advantage of accelerated amortization, and so any trends toward capital-saving technology or toward rounding out facilities that do not require necessarily heavy doses of capital outlay may well be responsive to the tax allowance incentive. In this, their conclusions suggest a set of "responses" over the cycle different from Chenery's.

These generalized empirical findings emphasize the need to

48. See for example the statements of steel executives referred to in Chapter 5.

49. *The Investment Decision: An Empirical Study*, pp. 16–17.

50. Ibid., p. 20.

51. Ibid., p. 109. This has clear implications for policy formation, and shall be referred to again in Chapter 6.

broaden the range of variables considered from that of the basically simple accelerator-centered model used in some of the material of Chapter 3. For purposes of assessing the industry's potential role in disrupting or sustaining levels of general economic activity, the limited model suggested sufficient (if not necessary) conditions, originating in the industry, that might have impact on the larger economy. As one moves from questions of industry impact on the economy to questions of motivational behavior within the industry, however, it is clear that the range of factors to be considered must extend beyond the simple linkage of expected increases in demand and anticipatory expansion of capacity, to such factors as the liquidity consideration, and to a number of other behavioral variables which will be developed more fully in Chapter 5. In analyzing the demand for the output of the steel industry itself, it follows that industry demand is subject to generalized investment pressures in the economy, consequences of the factors stressed by Meyer and Kuh. In their primary conclusions, they stress:

> (1) that the liquidity variables appear to be much more influential when over-all economic conditions are either stable or deflationary; (2) that a capacity formulation of the accelerator provides a reasonably good explanation of investment when the economy is under strong inflationary pressures and expanding rapidly; and (3) that, of the many possible subsidiary conditions that might determine a firm's sensitivity or insensitivity to capacity pressure, liquidity considerations appear to be the most important.[52]

They do qualify their conclusions with regard to differences between small and large firms, however. Small firms appear to be more sensitive to variations in liquidity position,[53] whereas larger firms have an investment policy linked more closely to capacity utilization. Meyer and Kuh suggest that this may be a reflection of the relative ability of firms of various sizes

52. Ibid., p. 117.
53. Ibid., p. 191.

to gain access to outside financing when the need is felt for it. Nevertheless, much of this behavior depends upon the personalities and inclinations of management in a given enterprise. Differences in behavior in resorting to outside financing may be as much a function of the particular individuals in charge of policy in a given firm as of the specific size of that firm.

One final study which relates to our discussion of previous attempts to analyze demand for the industry should be included here. It combines extrapolation techniques suggested by the very early attempts, but with a more sophisticated use of multiple regression. In its inclusion of various factors involved in anticipating demand, it recognizes influence consistent with the findings of Meyer and Kuh. This study, by Paul Boschan, stresses techniques used by the Econometric Institute in its approach to demand analysis.[54] Boschan divides his analysis into what he identifies as capacity-determined demand, that is, long-term demand for steel, and output-determined demand, a short-term component. He applies his data to the years 1919–40 and then tests them on the basis of extrapolating demand to the years around 1950. By correlating the Federal Reserve Index of Industrial Production with steel production figures, he finds a correlation coefficient of 0.897 over the period 1919–40.[55] He concludes that this simple relationship "does not provide a suitable basis for long-run projection." He proceeds to identify the long-term and short-term components of demand. In the course of his discussion (of interest here because of its relation to the bottleneck argument of Chapter 3) Boschan, by inverting one of his regression formulas and solving it for total industrial production, presents a formula which represents steel capacity as a limit to the total level of industrial production. He arrives at conclusions which suggest, in line with assertions of others who have studied this subject, that the price elasticity of demand for steel is relatively low.

> The area where substitutions are technologically or economically feasible comprises only a minor frac-

54. P. Boschan, "Productive Capacity, Industrial Production, and Steel Requirements," in *Long-Range Economic Projection*, pp. 233–58.
55. Ibid., p. 234.

> tion of aggregate steel demand. This being the case, how will the steel consumer who wants to stay in business react to an increase in steel prices? He will ask whether it is possible to shift the whole cost increase to his customers without impairing the size of the market for his product and he will consider whether his profits permit him to absorb the price increase. Only in a few cases will he find it possible to use a substitute or to reduce his unit demand for steel on short notice.[56]

Boschan emphasizes that a pertinent study of the demand for steel calls for a study of shifts in the demand curves themselves, not shifts along a given demand curve. The shifts themselves "will make shifts along the demand curve due to price changes appear unimportant." He points out that in an attempt to analyze the influences which cause shifts in the demand schedules themselves, the general level of industrial production can be examined, or can be broken down to the point of analyzing specific steel consumer behavior. This approach may be modified even further by introducing that phase of cyclical fluctuation in which the anticipated demand point is likely to occur. "The cyclical variations of the 'product-mix' of industrial production and variation in the 'working-inventory requirements' will be reflected in the short term component."[57] Beyond this, allowances for technological change which affect steel consumption patterns may also be made.

These suggestions lead our study to some comments on general criteria for effective demand analysis and then to presentation of one potentially fruitful demand analysis technique.

SOME CRITERIA FOR EFFECTIVE DEMAND ANALYSIS

In addition to the difficulties of using a simple accelerator, of ignoring rates of capacity utilization, or of turning one's back

56. Ibid., p. 256.
57. Idem.

on such questions as the influence of financial resources, there are also grounds for criticism in the absence of much-needed detail in more aggregative formulations such as Tinbergen's and those of Klein and Christ, to say nothing of Harrod and Domar. In commenting on this, for example, Duesenberry and Grosse argue:

> The reality underlying our models consists of a multitude of individual phenomena. We seek regular relationships among these phenomena so as to be able to predict one set of observations from another. Obviously we cannot work with the original phenomena. It would be useless as well as hopeless to try to explain every ring of the cash register. It is therefore necessary to classify and aggregate our observations in some way and rely on the laws of large numbers to bring regularity out of the chaos of individual observations. But if we aggregate too far we violate the very laws of large numbers on which we rely. We cannot suppose that the investment responses of two quite different industries to increases in their outputs are drawn from the same probability distribution. The same thing could be said of two subdivisions of the same industry or of two firms in the same subdivision. As we make our classification finer, however, the distributions in question become more homogeneous. These considerations would lead us in the direction of disaggregation even if we had to deal with a stationary system, i.e., one on which no exogenous changes impinge.[58]

Leontief has also stated the case for disaggregation, particularly as it impinges on investment processes. He states:

> Traditional capital theory, even more than other parts of abstract economic analysis, has suffered from unduly aggregative formulations. In the pres-

58. Duesenberry and Grosse, pp. 25–26.

> ent analysis the capital structure of the national economy is visualized as consisting of physically discernible stocks of specific commodities. The input-output ratios of the simple flow theory described above are supplemented by corresponding stock-flow relationships. These explicitly take account of the fundamental technological fact that the production of a flow of specific output requires not only the availability of corresponding flows of current inputs, but also the existence of previously accumulated stocks of equipment, buildings, inventories, intermediate products, etc. With the introduction of stock-flow relationships the system, of necessity, acquires dynamic character; the present rates of output become theoretically—as they certainly are actually—dependent upon the accumulation of the past input (investment) flows.[59]

Controversy regarding both of the issues raised in these paragraphs has been recurrent since the publication of Tinbergen's studies of 1939. Discussion of the relative merits of disaggregation cannot be elaborated further here;[60] however, we may consider this matter in the light of attempts at demand analysis for the steel industry itself, as reflected in the work of Paradiso, Dunn, et al., in the following way.

In their desire to come to grips with pressing policy ques-

59. Leontief, *Studies in the Structure of the American Economy*, p. 12.

60. Leontief, *The Structure of American Economy, 1919–1939*, p. 210: "It is easy to understand why aggregation and correlation go together. Direct observation can have only very limited use in discovery and explanation of quantitative interdependence of highly aggregative quantities. The very process of aggregation obscures the sharp outlines of the underlying structural relationships to such an extent that one is naturally forced to give up the simpler methods of direct induction and take recourse to 'blind flying' by the complicated but hardly foolproof instruments of indirect statistical inference." See also L. R. Klein, "Studies in Investment Behavior," in *Conference on Business Cycles* (New York, National Bureau of Economic Research, 1951) p. 233 ff.; W. Leontief, "Comment on 'Studies in Investment Behavior,'" ibid., pp. 311–313; and Klein, "Reply," ibid., p. 316. Leontief discusses a comparison of the errors of input-output and multiple regression methods of prediction in his *Progress Report, Preliminary* (Cambridge, Mass., Harvard Economic Research Project, 1952), section 5.

tions, both Dunn and Bean took paths to capacity estimates which led them over poorly defined ground. Setting aside evaluation based on only the relative pragmatic value of two approaches, mention should be made of certain significant factors, both favorable and unfavorable, in their methods.

Bean's approach is essentially one of trend projection of raw per capita data. Two major assumptions are implicit in trend projections of this type: (1) that the trend line is a good fit, and (2) that deviations from trend over time are not so significant as to invalidate the projection. The pitfalls suggested by proceeding on these assumptions lead one to restrict estimates to as short a projected time period as possible. Accuracy drops rapidly as estimates extend into the future. Bean, however, candidly talked in terms of 1960 and 1970, without making clear the tenuousness of the bases upon which his estimates rested.

To illustrate the use of another simple extrapolation technique, we might briefly return to the G.N.P.–steel production series used in Chapter 2. Two series, constant dollar G.N.P. and steel ingots, are set up, deriving regression equations for each. We find that deviations of each from their respective trends yield a coefficient of correlation of .832 (standard error of estimate of 8.95 million net tons). The data appear in Table 15. Next, we see what estimates of steel production would have been made in 1946, had the correct values for G.N.P. for the three following years been known. By substituting in the regression equation for the trend deviations of steel plotted against the trend deviations of G.N.P., estimates for steel are obtained.[61] They appear in Table 16 with the actual steel figures for the three years. The estimates prove to be off 5.8, 6.6, and 2.4 million net tons for the three years. Though this is consistent with the standard error of estimate of 8.95 million net tons, it involves error sufficiently great to make one wary of this type of quantitative estimation. The actual figures swing more widely than do the estimated figures, possibly due to the reason

61. An estimate of G.N.P. from known steel figures appears in Chapter 2, pp. 52 ff.

TABLE 15. *Data Concerning the Relationship between Movements in the Value of the Gross National Product and the Amount of U.S. Steel Production, 1919–46*

Year	*G.N.P. (billions of dollars)*[a]	*Steel ingots (millions of net tons)*[b]	*Trend for G.N.P.*[c]	*Trend for steel*[d]	*Deviations from trend, G.N.P.* x	*Deviations from trend, steel* y
1919	103.0	37.7	71.2	31.0	31.8	6.7
1920	97.1	45.8	76.4	32.5	20.7	13.3
1921	87.4	21.5	81.6	34.0	5.8	−12.5
1922	95.6	38.7	86.8	35.5	8.8	3.2
1923	109.6	48.7	92.0	37.0	17.6	11.7
1924	109.8	41.2	97.2	38.5	12.6	2.7
1925	114.7	49.4	102.4	40.0	12.3	9.4
1926	119.6	52.6	107.6	41.5	12.0	11.1
1927	119.8	49.0	112.8	43.0	7.0	6.0
1928	125.1	56.4	118.0	44.5	7.1	11.9
1929	133.0	61.4	123.2	46.0	9.8	15.4
1930	119.5	44.3	128.4	47.5	− 8.9	− 3.2
1931	109.6	28.5	133.6	49.0	−24.0	−20.5
1932	93.8	15.1	138.8	50.5	−45.0	−35.4
1933	94.8	25.6	144.0	52.0	−49.2	−26.4
1934	106.5	29.1	149.2	53.5	−42.7	−24.4
1935	115.5	38.0	154.4	55.0	−38.9	−17.0
1936	134.2	53.2	159.6	56.5	−25.4	− 3.3

TABLE 15—*Continued*

Year	*G.N.P. (billions of dollars)*[a]	*Steel ingots (millions of net tons)*[b]	*Trend for G.N.P.*[c]	*Trend for steel*[d]	*Deviations from trend, G.N.P.* x	*Deviations from trend, steel* y
1937	137.9	56.4	164.8	58.0	−26.9	− 1.6
1938	131.9	31.6	170.0	59.5	−38.1	−27.9
1939	142.8	52.5	175.2	61.0	−32.4	− 8.5
1940	157.5	66.6	180.4	62.5	−22.9	4.1
1941	186.8	82.4	185.6	64.0	1.2	18.4
1942	215.1	85.5	190.8	65.5	24.3	20.0
1943	244.6	88.4	196.0	67.0	48.6	21.4
1944	263.5	89.2	201.2	68.5	62.3	20.7
1945	260.6	79.4	206.4	70.0	54.2	9.4
1946	229.6	66.3	211.6	71.5	18.0	− 5.2

[a]Gross national production: United States Department of Commerce series converted to 1947 dollars.
[b]Steel ingot production from J. A. Krug, *National Resources and Foreign Aid*, Department of Commerce (Washington, Government Printing Office, October 9, 1947), p. 58.
[c]Regression equation for G.N.P.: $Y = 144.0 + 5.2X$; origin: 1933, X units = 1 year.
[d]Regression equation for steel: $Y = 52.0 + 1.5X$; origin: 1933, X units = 1 year.

Orcutt has suggested, namely that this situation might arise because the regression equation was calculated over a period dominated by a cyclical movement of different character or duration than that which might have been operative from 1947 to 1949;[62] hence, the inconsistent relationship between swings of the estimated and actual figures. It appears that one should be quite skeptical of this type of trend projection and associated estimations of deviations from trend—for the "cyclical character" reason as well as sectoral changes in technology, demand patterns, growth rates, and so forth.

TABLE 16. *Estimated and Actual Steel Production, 1947–49*

Year	*Estimated*	*Actual*
1947	78.82	84.6
1948	81.83	88.4
1949	80.19	77.8

Dunn, in contrast to Bean, did restrict himself to rather modest projections. His estimates, though based on cumulative past experience, were confined essentially to the year following his report; in this sense, he semed to be on more solid ground. Dunn assumd increases in capacity based on plant expansion plans "in the works," and from there, estimated adequacy in the next period based on what he assumed would be a continuing pattern of consumer demand (plus stated military needs). What he did not do was to allow for a re-orientation of the consumption pattern for steel in the future period. Bean did do this —by a straight line projection of per capita increases—but this could hardly be considered adequate treatment of what might be very complex change.

The work done by both these men is significant not only from the standpoint of particular contributions, but also because they emphasize the lack of completeness of modern analytical technique in coping with such problems as adequacy of steel capacity. In Bean's relating increases in steel demand not only to population increases, but to per capita trend as well,

62. Cf. p. 55 above.

we have the implicit suggestion that considerable qualification must be attached to judgments concerning consumption pattern changes. Dunn's approach, as an engineer, in proceeding by careful analysis of every stage in the productive process, suggests that the problem of bottlenecks in production as well as changes in the availability of input materials can and should be treated. In addition to this, his approach suggests that changes in quality and type of output demanded require ability to handle changes in finishing facilities requirements; that is, an improvement of method is needed, beyond mere inspection of ingot production as a sufficient reflection of capacity. Dunn's approach points to the desirability of building an analytical structure that will permit adjustment of production estimates when, through technological change, changes occur in the production coefficients.

Dunn's approach combines, in a sense, the use of historical series and of projections of trend based on these series, with an empirical survey of industry conditions by a man of sound engineering knowledge. The result might well be considered an amalgam of the best data available for establishing the most logical basis for estimating productive potential. On the other hand, it could be claimed that lack of internal consistency is a telling weakness of his approach. Although Dunn was forced, by the practical contingencies of his assignment, to accept and incorporate the best available data into his analysis, he was, nevertheless, the victim of an array of incomplete considerations. For example, his effective employment of technological information is certainly important and relevant in evaluating his analytical work. But his lack of a model, which might have afforded him a chance to see the interaction of changes in one part of the system with changes in other parts or to recognize changes from the outside of an industry altering needs in the internal structure, forced him into an analytical situation susceptible to breakdown with the slightest alteration of external conditions.[63]

63. In a post-war reference to Dunn's efforts, Senator O'Mahoney commented: "[his] report was to the effect that the steel capacity of the United

In contrast to this, let us consider the desirability of a system which does not necessarily deliver the investigator into a forced set of ceteris paribus assumptions; rather, it allows him to see the interaction of the industry and the economy; its very formulation makes allowance for all the interacting variables. This is an abstract way to say that a model is needed which allows for the multiple character of the demand for steel. This demand has its origin in six principal sources: for use in current output of consumers' durable goods, in output of capital equipment for replacement,[64] capital goods for new capital formation, in output for additions to steel capacity, for export, and for special demands such as armaments. Each of these can be said to be dependent on factors moving at different rates; for example, demand for steel going to consumers' durables will have a more remote relationship to the amount of investment in the economy than will the demand for steel to be channeled to the enlargement of steel capacity—the latter taking place in some cases when very high levels of net investment are expected. A model that can account for this complex demand picture and the dynamic character of the consumption pattern for steel can be expected to be far superior to the methods we have thus far described.

States was sufficient for all foreseeable purposes in the war. I don't think Mr. Dunn was to be criticized for that report because the estimates which were being furnished by the Army and Navy at that time are probably justified in the report. It was not until much later that the Army and Navy realized to what extent we would have to expand production, and it was expanded by the investment of Government funds." U.S. Congress, Senate, "War Plant Disposal—Iron and Steel Plants," Sub-Committee on Surplus Property of the Committee on Military Affairs and the Industrial Reorganization Sub-Committee of the Special Committee on Post-War Economic Policy and Planning (Washington, D.C., Government Printing Office, 1946), p. 3.

64. For a treatment of capital goods requirements for replacement needs, in the analytical context of material below, see R. N. Grosse and E. B. Berman, "Estimating of Future Purchases of Capital Equipment for Replacement," in *Problems of Capital Formation,* A Report of the Conference on Research in Income and Wealth, National Bureau of Economic Research, (Princeton, Princeton University Press, 1957), pp. 389–415.

THE INTERINDUSTRY FORMULATION RELATED TO ANALYSIS OF DEMAND FOR STEEL

In addition to an analytical framework which will allow sensitive and detailed measurement of the multiple sources of demand for the output of the industry, it is also desirable to find a structure which will introduce the effects of changes in the level of national income, changes in requirements for steel in the economy outside of the industry due to changing technology or because of substitution, changes in the availability of input factors to the industry, changes in the production function of the industry, changes in factor use within the industry via substitution, and finally changes in the product mix of the industry. We wish to measure the effect of these changes on the industry, and in the over-all economy as well.

It will be suggested in subsequent paragraphs that the traditional Leontief input-output matrix provides one means for discovering the specific variables affecting the steel industry, and at the same time allows for inclusion of the various considerations listed above.

Before proceeding with a discussion of its merits, however, we should be aware that methodological problems do exist. Input-output techniques do not relieve us of statistical problems and do not provide us with a completely satisfactory descriptive tool. In Koopman's words:

> It may be useful to draw a distinction between planning models designed to study and guide optimum allocation of resources toward some stated objective, and models intended as a description of economic behavior in reality. Reference to this distinction will clarify apparent contradictions. For instance, the acceleration principle, declared dead yesterday by Professor Haberler, is very much alive today in Professor Leontief's presentation. The contradiction is resolved if we realize that the accel-

> eration principle is dead as an explanation of actual investment fluctuations during business cycles, but makes very good sense in a model designed to plan investments over a period in order to attain levels of production increasing in a pre-assigned manner. Similarly, the use of technological information stressed by Leontief seems adequate for a planning model, but the attempt to develop a dynamic model of actual behavior including such matters as investment in inventories and in productive equipment is likely to reintroduce the difficulties of statistical inference from time series which he has been hoping to avoid.[65]

Nevertheless, as a demand analysis instrument, it has significance for us in its pragmatic potentialities.[66]

The development of the interindustry approach has its origins, in a generic sense, in the theoretical work of Walras,[67] but more specifically in the work of Leontief since the 1930s.[68]

65. T. C. Koopmans, "Comment" on W. Leontief, "Input-Output Analysis and its Use in Peace and War Economies: Recent Developments in the Study of Interindustrial Relationships," *American Economic Review, 34 (3)* (1949), 234.

66. E.g., A. Grosse, "Textile Production Functions, Equipment Requirements, and Technological Change, A Study in the Utilization of Primary Technical Information," (Harvard Economic Research Project, December 1946); Leontief, *Studies in the Structure of the American Economy*, parts 2–5; Duesenberry and Grosse, "Technological Change and Dynamic Models"; S. S. Netrebs, "The Development of the Bill of Goods for Interindustry Analysis," Conference on Research in Income and Wealth (New York, National Bureau of Economic Research, 1952); E. Glaser, "Inter-industry Economics Research Program"; A. Phillips, "Plant Capacity: a Study in the Non-Friction Bearing Industry" (unpublished Ph.D. dissertation, Dept. of Economics, Harvard University, 1953); Grosse and Berman, "Estimation of Future Purchases."

67. For a discussion of Walras, see G. Stigler, *Production and Distribution Theories* (New York, Macmillan, 1949), chap. 9; for an analogy in matrix form of the Walras-Leontief approach, see R. M. Goodwin, *Proceedings of a Conference on Inter-Industrial Relations* (Driebergen, Holland, 1958).

68. W. Leontief, "Quantitative Input and Output Relations in the Economic System of the United States," *Review of Economic Statistics, 18* (1936), 105–25; "Interrelations of Prices, Output, Savings, and Investment," *Review of Economic Statistics, 19* (1937), 109–32; "Output, Employment, Consumption, and Investment," *Quarterly Journal of Economics, 58* (1944), 290–314; "The Economics of Industrial Interdependence," *Dun's Review*

In commenting on distinguishing features of the input-output approach, Evans and Hoffenberg point out that interrelationships are therein specified and applied to defined sectors. They add that in contrast to "indefinite" or general correlation, these relationships are consistent with "established custom or technological necessity."[69] They add that because detail is not obscured in broad aggregates, "specific questions may be put to the analytical machinery and equally detailed results for its sectors may be obtained." And they also note the flexibility and wide applicability of the approach, indicating for example that "changes in the demands made on the economy or changes in its processing structure may be stipulated as desired and the consequences examined."

In the use of a general equilibrium framework, the interindustry approach employs a rectangular input-output table to characterize the flow of commodities from each industry in the economy to all the respective related industries. The activities included in this table are not restricted to production, but also cover distribution, consumption, transportation, foreign trade, government services, and so forth. The various transfers of goods and services among sectors, within a specified period of time, are set forth in such a manner that the total output of a particular industry or sector is reflected in the row for that sector and the distribution of all inputs flowing to a given sector is listed in the column under that sector designation. In this form, a table with an equal number of rows and columns is presented reflecting the disaggregated outputs of the economy and at the same time showing to what specific consumers each

(February, 1946), p. 22; "Computational Problems Arising in Connection with Economic Analysis of Industrial Relationships," *Proceedings of a Symposium on Large-Scale Digital Calculating Machinery* (Cambridge, Mass., Harvard University Press, 1948), p. 169; "Recent Developments in the Study of Interindustrial Relations," *American Economic Review, 39* (1949), 211–25; *The Structure of the American Economy, 1919–1939;* "Some Basic Problems of Empirical Input-Output," (National Bureau of Economic Research, 1952); *Studies in the Structure of the American Economy,* chaps. 1–3.

69. Evans and Hoffenberg, "The Interindustry Relations Study for 1947," p. 127.

given industry's output flows. The equilibrium of the total set of relationships is reflected in the fundamental equation:

$$X_i - \sum_{k=1}^{m} x_{ik} = y_i \qquad i = 1, 2, \ldots, m$$

where X_i represents the annual rate of total output of industry i; x_{ik} represents the amount of product of industry i absorbed annually by industry k; and y_i represents the amount of the same product i made available for consumption by some consuming unit other than one of the m industries covered within the system under consideration.[70] Over-all balance is expressed in m linear equations under the assumption that there are m industries in the system.

The structural relationship between the various inputs responsible for given output (or conversely the output forthcoming as a result of a roster of inputs) is expressed in the following relationships:

$$x_{ik} = a_{ik} X_k \qquad \begin{array}{l} i = 1, 2, \ldots, m \\ k = 1, 2, \ldots, m \end{array}$$

where a_{ik} indicates the amount of each particular input used by that industry per unit of its own output. Substituting from above:

$$X_i - \sum_{k=1}^{m} a_{ik} X_k = y_i \qquad i = 1, 2, \ldots, m$$

When this system is solved for the total output of the given industries, the m linear equations (with m unknowns) will take the form:

$$X_i = \sum_{k=1}^{m} A_{ik} y_k \qquad i = 1, 2, \ldots, m$$

70. Leontief, *Studies in the Structure of the American Economy*, p. 18.

In this context, when a final bill of goods is formulated or "decided upon," the rate of output of industry i can be determined. A matrix of the technical coefficients—that is, the matrix:

$$a = \begin{bmatrix} a_{11} & a_{12} & \cdots & a_{1m} \\ a_{21} & a_{22} & \cdots & a_{2m} \\ \cdot & \cdot & & \cdot \\ \cdot & \cdot & & \cdot \\ \cdot & \cdot & & \cdot \\ a_{m1} & a_{m2} & \cdots & a_{mm} \end{bmatrix}$$

represents in each column the input coefficients of each given industry and furthermore, "Each of the constants, A_{ik}, is in general a function of all the a's, i.e., it depends on the input coefficients of all the m industries."[71]

Objections to this approach have been raised because of concern over the relative rigidity of proportions indicated by the flow coefficients and also their relative fixity over time. In regard to the first of these, Leontief has noted:

> Insofar as the proportion in which the separate factors can be combined within the same production function . . . are variable, these proportions will most probably vary with every change in their relative prices. . . . It is, however, not the fundamental validity of the principle of substitution but its quantitative significance which is important from the point of view of empirical analysis. The smaller the variation in production coefficients induced by any given range of changes in factor prices, the smaller obviously will be the empirical error introduced in our computations by the assumption of invariable input ratios. . . . An indirect test presented elsewhere [see A and B in part 4 of *The Structure of American Economy, 1919–1939*] seems to indicate that, within

71. Idem.

> the range of price changes which actually have taken place between 1929 and 1939, these errors lie within relatively narrow limits. . . .
>
> For purposes of prognostication the possibility of large spontaneous changes in the magnitude of technical input coefficients would, of course, be as damaging as any secondary change of comparable magnitude induced through price variations. Here again the indirect test referred to above seems to justify expectation of a considerable degree of stability in the general pattern of price relationships.[72]

More specifically related to conditions affecting the steel industry, the relative fixity represented by the technological coefficient may be said to approach the realities of the situation more closely than "a range of substitution." For example, in the case of aluminum, although acceptable in quality, if it still will not be widely substituted for steel because steel will retain its cost advantages over a considerable range, then only when the relative prices reach some critical point will a threshold of substitution be passed. There are barriers to substitution, of course, also because of consumer resistance, e.g., to the use of plastic in place of steel because it does not "appear" as strong, and other rigidities in consumer preference; because of already achieved economies of production and vested interest in already existing plant facilities and resource access and hence oligopolistic pressures to maintain the market. In regard to this, Leontief has observed: "In all the numerous cases in which every cost factor has an unlimited range of possible substitutes, the actual price variations are safely removed from such critical points, so that the assumption of constant pro-

72. Leontief, *The Structure of American Economy, 1919–1939,* pp. 201–02. There has been some tendency to regard these two methods (input-output technique versus multiple regression analysis) as competitive. Continued study of the problem, however, indicates rather that they are complementary. Each method has its own relative merits; input-output for its recognition of the stable structural relationships in production, multiple regression for its explicit inclusion of a "growth" factor which comprises the unstable elements through time. Indeed, it is possible that the combination of the two methods might result in better predictions than the use of one alone.

portions is for all practical purposes entirely justified."[73] It should be added that though relative stability of the production functions of the consumers of steel (as well as within the industry) is assumed, there is no preconception regarding the range of output flowing to the final bill of goods. Therefore, the sequence of indirect effect from the ultimate level of output in the economy to the industry is sustained and embodied in the interindustry approach.[74]

Not only is this analytical mechanism capable of satisfying the requirements for reflection of other numerous categories of changes referred to above, but it also lends itself to flexible use as far as degree of disaggregation is concerned. This is true, for example, in terms of dividing a given sector, such as capital goods, into the various industries that might be classified in this category. It is also true at a more detailed level, in that specific industries, such as steel, can be broken down into their component parts. The limit to this disaggregation can be said to be set by the character of the available data; it also may be set by the capacities of the computational devices available for the analysis.[75] In spite of this, however, the theoretical formulation does not impose specific limitations to disaggregation and it may be expected that computational and data problems will ultimately be solved. Moreover, it permits considerable disaggregation in one sector at the same time that information concerning other sectors may be treated in more aggregative form. This is of particular significance for the proposed device which is to follow.[76]

Assume that the special concern of the analyst is to obtain quantitative appraisal of a prospective demand situation to be

73. Leontief, *The Structure of American Economy, 1919–1939*, p. 41.

74. Elaboration to include stock-flow relationships and the problem of structural change will be treated below, pp. 152–57.

75. Certain general theoretical problems faced in quantitative demand analysis, including statistical and mathematical problems, time dimension choice, and limitations of empirical materials, cannot be discussed here; it is true, however, that until more completely resolved, they may set limits on certain present-day uses of the approach.

76. "First, many practical problems are concerned with industrial detail in its own right. Hence, in questioning the adequacy of industrial capacity to supply some combination of military and civilian purchases we cannot throw

faced by the steel industry in the context of given structural relationships and subject to a determined final bill of goods. Assume further that it is of considerable importance to convey, in this evaluation, the situation of the steel industry vis-à-vis the total economy, because it will be used by steel entrepreneurs preoccupied with their own position in terms of investment decisions. With these objectives, then, it can be suggested that the steel industry, as a sector, be extracted from the input-output matrix in the conventional formulation of interindustry relationships, and that it be placed outside of the system in its entirety.[77] This is symbolized by Figure 9 where, in effect, the steel industry is treated in much the same way as foreign trade can be treated. Here the economy "exports" input factors to the steel industry, and the output of the steel industry is "imported" by the economy.

Up to this point, there has not been an attempt to pin down the exact limits of what may be classified as the iron and steel

---

together all manufacturing production. As successive splits are made in the classification-aggregation system, metals industries can be isolated, then metal smelting, then steel production, then open hearth steel, and so on with increasingly useful results. These refinements are generally accompanied by increasing difficulty of obtaining usable data, and increasingly difficult computation problems. Secondly, these tables express average or 'normal' concepts of structural relationship, which may be more satisfactorily dealt with in more detailed layouts. For example, the inherent assumption of the simplest models is that the output of each industry has a stable product-mix (or has a set of compensating changes that preserves the validity of its input coefficients). The acceptability of such an assumption can better be analyzed for detailed relatively-homogeneous industries than for some large heterogeneous aggregate. Moreover, if there is reason to specify changes in these structural relationships, it can be done better by specifying exactly where in the industrial processing these technological changes are postulated, rather than to dilute them in large statistical aggregates." Glaser, "Inter-industry Economics," pp. 6–7.

77. By enumerating a list of products and by stating that the term "member of the industry" means and includes companies which produce such products, the Iron and Steel Code briefly disposed of the difficult problem of industrial classification. An examination of the Code products discloses, however, that this definition brings into the iron and steel industry any company, no matter what its finished product, which by reason of integrated operations produces its own steel or iron as a raw material for further processing. Hence, by this definition the Ford Motor Company and the International Harvester Company, for example, were classified as members of the iron and steel industry. In fact, the heterogeneity of products and the differences in degree of integration make it impossible to segregate into distinct,

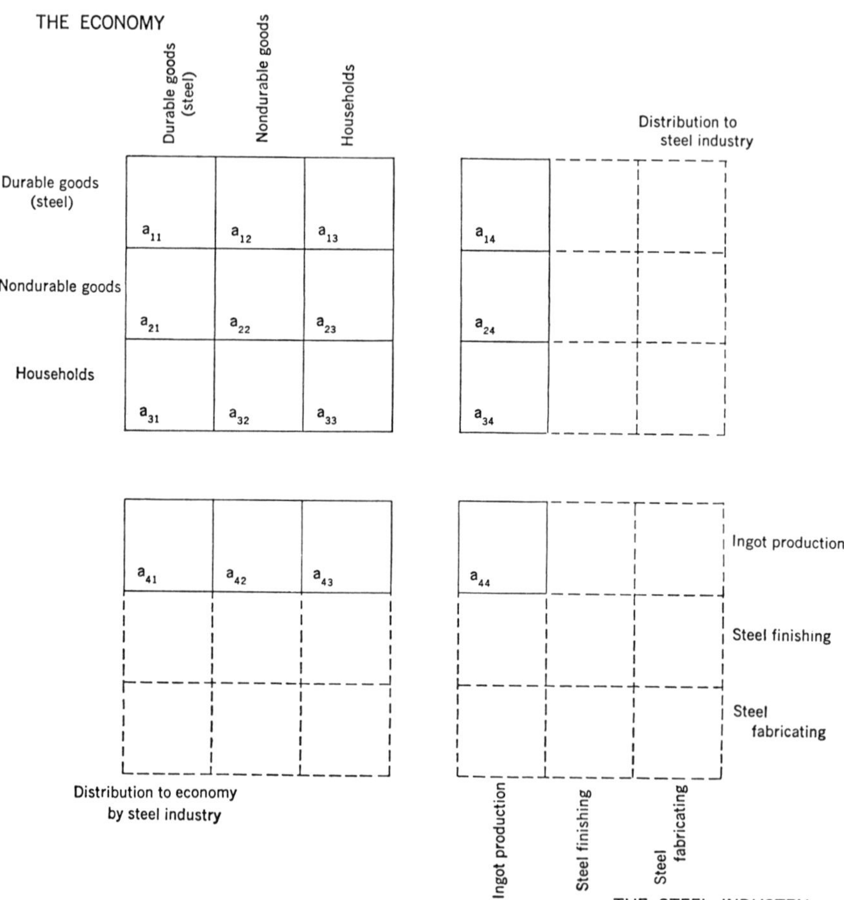

Figure 9. Separation of Steel Sector from Interindustry Structure

industry.[78] Companies engaged in the production of pig iron, ingots, castings, rolled products, ferrous alloys all are considered, under usual conditions, to be a part of the steel industry.[79] So also have been certain coke producers, mine operators, and

---

non-overlapping groups the producers of iron and steel (the raw material) and the manufacturers of fabricated iron and steel products. It should be pointed out, however, that in the iron and steel industry it was not necessary to formulate precise criteria of industrial classification for the purposes of Code organization. "In the membership of the American Iron and Steel Institute founded in 1908, was to be found the group of enterprises commonly referred to as the iron and steel industry. It was common economic interests, combined with similarity of economic characteristics, which caused this group to become segregated from the miscellany of firms engaged in some manner in the manufacture of iron and steel products. These companies, which for the most part became members of the Code, may be divided into two main categories: (1) producers of only finished rolled products—the non-integrated group; (2) producers of both finished rolled products and the crude and semi-finished products from which the finished products are processed. These members of the Institute are, in turn, differentiated from a larger and more heterogeneous group since they do not, characteristically, fabricate iron and steel, but sell it to manufacturers of metal products as a raw material for further processing." Daugherty, de Chazeau, and Stratton, *Economics of the Iron and Steel Industry, 1*, 9–10.

78. No specific classification of capital goods industries as such has been presented. It has been felt that in essence there is a relatively wide spectrum of borderline industries, and that any classification would of necessity be rather arbitrary. In addition, there is some question as to whether a more meaningful classification might not be those industries which are the large consumers of capital goods rather than those which are the large producers of capital goods. However, with regard to the latter criterion, when one surveys the ultimate percentage disposition of gross output to final demand category, it can be seen that such industries as agriculture and fish, food, tobacco products, textiles, and apparel can be considered as consumer goods industries on the basis that 10 per cent or less of gross output went into investment. On the other hand, such industries as iron and steel, non-ferrous metals, agriculture and construction machinery, and motors and generators may be considered capital goods industries on the basis that 40 per cent or more of gross output went into investment. (Figures are for 1947, and as the data do not include output directed to government or exports, it can be assumed that the so-called investment industries might have directed more nearly 50 to 60 per cent of their output into investment.) Cf. Evans and Hoffenberg, "Interindustry Relations Study," p. 123.

79. A detailed discussion of the classification problem specifically associated with the use for determining flow coefficients for the iron and steel industry has been undertaken by the Office of Business Economics, Department of Commerce. Of particular interest is the comparison between the approach of the Office of Business Economics and previous attempts at classification of the industry by the Bureau of Labor Statistics. Whereas O.B.E. and B.L.S. were in agreement as regards flow of ingots, intra-plant transfers for use in maintenance, shipments to other companies and interplant transfers within

railroad and steamship undertakings. United States Steel Corporation, for example, is a large producer of portland cement; Bethlehem Steel has built railway cars. Some steel producing operations extend straight through from the mining of the ore to the fabricating of the finished steel; others perform only one stage of the production process. One possible approach to delimiting the industry is on the basis of so-called common economic interests coupled to similar production characteristics. On the other hand, the category might be set at some point such as producers of only finished rolled products, or producers of both finished rolled products and crude and semi-finished products (integrated).[80] If the objective of classification is such that the demand faced by a given firm as well as the industry as a whole is desired, then questions of market share must be combined with total industry figures. And total industry limits must be set so that *balanced* flow in response to consumer demand is represented. In practical terms, industry output can be taken

the steel industry for use in rolled or fabricated products, shipments and interplant transfers to non-steel industries; and where they were in agreement on flow of finished steel products, intra-plant transfers for use in rolled products and for use in maintenance, also shipments to other companies and interplant transfers within the steel industry for use in fabricated products and also shipments and interplant transfers to non-steel industries; the two agencies were in disagreement on the flow of ingots, interplant transfers for use in rolled products or fabricated products, where O.B.E. considered them input flows and B.L.S. output flows; also under shipments to other companies and interplant transfers within the steel industry for use in maintenance, where O.B.E. considered these outflows and B.L.S. inflows; also in flow of finished steel products, intra-plant transfers for use in fabricated products, where O.B.E. considered these inflows and B.L.S. outflows; and finally shipments to other companies and interplant transfers within the steel industry for use in rolled products and for use in maintenance, where O.B.E. considered these outflows and B.L.S. considered these inflows. Cf. "Steel Ingots and Finished Steel: Flow Inputs and Outputs," Office of Business Economics, Industrial Capacity Section, U.S. Department of Commerce, under general supervision of Murray Foss (processed). As indicated in the report: "The definition of output adopted here introduces certain flows which either do not show up in the Bureau of Labor Statistics shipments or transfers, or do so only to a very limited extent. Virtually all ingots consumed in the rolling process are consumed in the establishment making the ingots. In addition, most of the steel consumed in fabricating departments of steel mills does not show up as a Bureau of Labor Statistics shipment or transfer but may be considered instead an intraplant transfer." Ibid., p. 6.

80. Daugherty, de Chazeau, and Stratton, *Economics of the Iron and Steel Industry, 1,* 10.

as that aggregate included in the data of the American Iron and Steel Institute.[81]

As mentioned above, both matrices—that of steel and of the over-all economy—can be broken down into precisely those stages of production (in the case of steel) and those industries which provide steel with its input, to give the investigator a mechanism which provides sensitive reflection of structural characteristics and demand conditions, and which provides openings to make adjustments in the over-all structure necessary because of changes in parameters. For example, it might be analytically desirable to split the general economy matrix into an asymmetrical roster of categories such as consumers goods, producers goods, government and households, coal, iron ore, transportation, non-ferrous mining, limestone, scrap, and machinery. At the same time, it might be desirable to break down the steel matrix (represented in the diagram above by the broken lines, suggesting the four by four matrix expanded to six by six, and so forth) into such categories as coking facilities, blast furnace, open hearth, electric furnace, finishing mills, and fabrication. This array would bring to the surface the particularly relevant sectors or factors. If, for example, technological innovation causes a change in pig iron production, adjustment can be made of the flow (or capital coefficient) in the blast furnace sector, and then the effects can be traced throughout.[82]

81. Recall that there is room for confusion as regards the significance of capacity figures; see pages 94, 124. For example, the American Iron and Steel Institute decided at one point to establish a new index for steel output because of time discrepancies in the computation of capacity figures and the need for useful operating rate figures. Cf. *The New York Times* (December 28, 1953) p. 29. Age distribution of existing capacity is another relevant consideration; also the fact that standby capacity may be brought into use for some purposes during peak load periods but for other purposes it may be technologically inappropriate or incapable of providing sustained output for the duration of the peak load.

82. "For example, steel is typically distributed mainly to other processing industries, and it usually passes through many hands before reaching the purchaser of a finished commodity, such as an automobile, a can of food, and so on. Yet it is the demand for specific finished goods that determines the market for steel. Using interindustry relations methods, steel industry shipments may be traced forward to show both dependence on consumer, government, investor, or export demand for food products, furniture, new houses, automobiles, and so forth. Similar quantitative expressions of the relationship

It is readily apparent, particularly in light of the material developed in Chapter 3, that the role of the steel industry is primarily related to investment activity, and stock-flow relationships can be better expected to be a mandatory part of a demand analysis framework for steel than for perhaps less investment-oriented industries. The interindustry approach and the corresponding matrix extraction lends itself effectively to this problem.[83]

The fundamental equation cited on page 143 can be written to include the stock-flow relationship. $S_{ik}(t)$ stands for the stock of a commodity produced by industry i and used by industry k at time t.[84] $ds_{ik}(t)/dt$ or $\dot{S}_{ik}$, then represents the rate of increase or decrease in that stock at time t, and the balance equation becomes:[85]

$$X_i - \sum_{k=1}^{m} x_{ik} - \sum_{k=1}^{m} \dot{S}_{ik} = Y_i \qquad \begin{matrix} i = 1, 2, \ldots, m \\ k = 1, 2, \ldots, m \end{matrix}$$

The structural relationships described on page 143 must now be joined by a set of relationships reflecting the capital output structure.

$$S_{ik} = b_{ik} X_k \qquad \begin{matrix} i = 1, 2, \ldots, m \\ k = 1, 2, \ldots, m \end{matrix}$$

To transform the structural equations into a form expressing the changes in stock of given industries, $\dot{S}_{ik}$, related to changes in the rates of output of these industries, $\dot{X}_{ik}$, differentiation in

between production and the ultimate demands of the economy can be carried through for any sector, and for this reason the approach constitutes an important and powerful new tool for business marketing research" (Evans and Hoffenberg, "Interindustry Relations Study").

83. Cf. K. J. Arrow and M. Hoffenberg, with the assistance of H. Markowitz and R. Shephard, *A Time Series Analysis of Inter Industry Demand* (Amsterdam, North Holland Publishing Co., 1959). This is a study, using a fixed coefficients interindustry model, which stresses the interrelated demand patterns of various industries.

84. Leontief, *Studies in the Structure of the American Economy,* p. 56.

85. The output of the steel industry will be controlled by the second left

respect to time on both sides of the equation must be undertaken:

$$\dot{S}_{ik} = b_{ik}\dot{X}_k \qquad \begin{array}{l} i = 1, 2, \ldots, m \\ k = 1, 2, \ldots, m \end{array}$$

At this point the additional set of structural relationships can be added to the system of equations on page 143, thus yielding:

$$X_i - \sum_{k=1}^{m} a_{ik}X_k - \sum_{k=1}^{m} b_{ik}\dot{X}_k = Y_i \qquad \begin{array}{l} i = 1, 2, \ldots, m \\ k = 1, 2, \ldots, m \end{array}$$

providing the final dynamic input-output formulation. The interaction in the system not only encompasses flows on current account but also the effects of rates of change of these flows. The corresponding set of capital coefficients can be expressed in the matrix similar to that of the flow coefficients:

$$b = \begin{bmatrix} b_{11} & b_{12} & \ldots & b_{1m} \\ b_{21} & b_{22} & \ldots & b_{2m} \\ \cdot & \cdot & & \cdot \\ \cdot & \cdot & & \cdot \\ \cdot & \cdot & & \cdot \\ b_{m1} & b_{m2} & \ldots & b_{mm} \end{bmatrix}$$

Because of the inclusion of both the flow relations and inputs absorbed on capital account, Leontief summarizes the significance of the dynamic system as leading to:

> the determination, or perhaps prediction, of the behavior of each one of these variables over time. Once the time shape of a particular output function, $X_k(t)$

---

hand term as well as the stock-flow term, as "all allocations of commodity i to current replacement and maintenance requirements of the capital goods and other stocks are to be thought of as being accounted for by the second term, $\sum_{k=1}^{m} x_{ik}$, unless of course they are included in the final demand, Y_i," Leontief, *Studies in the Structure of the American Economy*, p. 56.

has been determined, the corresponding variations in the input flows, $x_{ik}(t)$, absorbed and capital stocks, $S_{ik}(t)$, held in this k^{th} sector of the economy can be found by substituting $X_k(t)$ in the appropriate equations.[86]

Probably the most complete empirical study of investment costs associated with the production of new capital facilities, with the ultimate objective of deriving capital coefficients for the iron and steel industry, was undertaken by the Office of Business Economics for the Department of the Air Force in 1953. In addition to presenting a discussion of data and classification problems and suggested values for capital coefficients in the industry, such questions as interplant variations in capital coefficients, capital input lead times, and measurement of capacity problems are discussed.[87] Although it is beyond the

86. Ibid., p. 57. For a discussion of irreversibilities due to non-transferability of stocks and to technical limits to wastage or scrappage, see ibid., pp. 68–69. Also, for a discussion of structural change and its relationship to the formulation of a dynamic system see ibid., pp. 20–22. Leontief's *Structure of American Economy, 1919–1939,* pp. 63–64, contains a discussion of the relationship of productivity changes to the system. Duesenberry and Grosse also treat the causes for changes in outputs and the effect of technical change as well as the basis on which judgment may be made regarding the speed with which coefficients change, cf. "Technological Change and Dynamic Models," pp. 6–7, 82–83, and 96. For example: "Investment in each industry takes place in response to changes in its rate of output and in available technical alternatives and other exogenous events. Changes in outputs may result from autonomous changes in final demand, or from changes in input coefficients of other industries, or directly from changes in the rates of investment themselves. In addition, the investment of one period is a determinant of the output and therefore the investment of the next. Thus technical change has two kinds of influence on the process of development. On the one hand, when it is made effective through investment, invention changes input coefficients and thereby changes the production potential of the economy's resources. On the other hand, by changing input coefficients and by the effect of investment on final demand, invention affects the rate of utilization of resources. These considerations dictate the broad outlines of our model. Two relationships are emphasized—the relation between technical change, changes in output, and investment; and the relation between investment and changes in the flow coefficients" (pp. 6–7).

87. Cf. U.S. Department of Commerce, Office of Business Economics, "Investment Cost and Capacity in Iron and Steel: An Exploratory Study," prepared by the Industrial Capacity Section, Business Structure Division for the Interindustry Economic Research Program of the U.S. Department of the Air Force, September, 1953 (processed).

scope of this study to include detailed findings of the Office of Business Economics, it may be useful to illustrate the range of coefficients apparent in various stages of the production process by reproducing here part of the summary table from their study (Table 17). Though the Office of Business Economics study

TABLE 17. *Investment Costs in 1951 Dollars per Ton of Added Capacity, Based on Sample Expansions*[a]

Stage	*Unit of capacity measurement*	*Cost per ton (1951 dollars)*
Blast furnaces	Pig iron	36.70
Coke ovens	Coke	31.77
Open hearth furnaces	Ingots	32.53
Electric furnaces	Ingots	17.21
Slabbing and blooming mills	Slabs and blooms	16.80
Plate mills	Plates	52.26
Structural mills	Structural shapes	80.55
Bar and rod mills	Bars and rods	44.55
General plant		22.1%[b]
Steel foundries, open hearth	Finished steel castings	240.28[c]
Steel foundries, electric furnace	Finished steel castings	274.79[c]
Malleable iron foundries	Finished malleable castings	280.10[c]

[a]From Table 1, U.S. Department of Commerce, Office of Business Economics, "Investment Cost and Capacity in Iron and Steel: An Exploratory Study," p. 22. Costs are based on sample expansions.

[b]Per cent of cost of integrated steel plant.

[c]1947 dollars.

represents an impressive attempt to develop materials needed for interindustry analysis, it nevertheless does contain certain weaknesses. In the introduction to the study, for example, the following cautionary statement should be noted:

> Although considerable time was devoted to the derivation of the detailed data presented here they must

> be considered provisional because of the limited samples on which they are based. In particular, the figures on investment costs of specific materials and equipment by industry of origin must be used with caution since in many instances the data showed substantial variation from case to case. Because of larger samples and lesser variability the total cost figures are more accurate than the detailed statistics.[88]

In spite of this, however, the report does represent an advance toward shaping empirical materials in a fashion appropriate to the needs of this analytical approach, and it serves as a good case study of useful effort in this area. This is particularly true because of its range of coverage. It includes presentation of detailed capital coefficients, "description of the facilities and methodological notes for 'plants' embracing each of the following: coke ovens, blast furnaces, open hearth furnaces, electric arc melting furnaces, slabbing and blooming mills, plate mills, bar mills, structural mills, and iron and steel foundries."[89]

To relate the foregoing formulation of flow and stock coefficients to some of the variables discussed in Chapter 3, and to reflect how these coefficients may be keyed to investment changes of the so-called "autonomous type" as differentiated from that which is induced, the following matrix-equation is suggested:

$$[a]\begin{Bmatrix} r_1(t) \\ r_n(t) \end{Bmatrix} + \begin{Bmatrix} G_1(t) \\ G_n(t) \end{Bmatrix} + [b]\begin{bmatrix} r_1(t)-r_1(t-1) \\ r_n(t)-r_n(t-1) \end{bmatrix} = \begin{Bmatrix} r_1(t-1) \\ r_n(t-1) \end{Bmatrix}$$

where the a matrix is of flow coefficients, r stands for revenues (gross), G for such factors as autonomous investment and certain mandatory fixed capital; the b matrix is of capital coefficients and, taken with changes in revenue over the preceding

88. Ibid., pp. 3–4.
89. Ibid., p. 3.

period, accounts for induced investment.[90] In this equation we have the implicit synthesis of interindustry relationships with changing rates of income (output) in the economy and with varying magnitudes of investment (autonomous, induced, and replacement, if these distinctions can be said to have significance). More than this, the matrices of flow (a) and capital (b) coefficients which appear are subject to the type of breakup and extraction referred to above (pages 146–48 and 151) and are therefore especially appropriate to satisfy the need for an apparatus for demand analysis characterized earlier in this chapter. As theoretical formulation and empirical work proceeds,[91] a structure may be expected to evolve, capable of quantitative treatment of interaction between sectors of the economy over time. As it develops, the Leontief system may not only permit tracing of demand on specific industries as a result of changes in the final bill of goods, but accomplishing this within the context of a changing economic structure and the dynamic process.[92]

What is obscured perhaps by the over-all value of analyzing cumulative effects of such changes is the importance a structure such as this has for demand analysis from the standpoint of the individual industry or perhaps the individual firm. Neither the use of the matrix extraction device nor the equation above involves, of course, a modification of the basic theoretical formulation. It is presented, however, as a possible modification of form which would make the potentialities of

90. The choice of lag intervals can be very critical, the most desirable being one where the interval used has relevance to the policy objectives involved. For example, if the repercussions of some event such as war emergency, or, in a somewhat different sense, the avoidance of a bottleneck condition in the short run were to be traced, appropriate lag intervals (lead times) would be desired. (Cf. Glaser, "Inter-industry Economics," pp. 7–10.) Reference here is not only to gestation lag, but also to revenue lag, decision lag, payment of factor interval, dissipation interval, among others.

91. In addition to the need for data on the timing of phases of production, information will be needed concerning capacity utilization, age distribution of equipment, best-practice operation vis-à-vis normative conditions, and treatment of inventories, among other considerations.

92. See pp. 61–62.

the approach more accessible—perhaps more palatable—to those within industry who are responsible for investment decisions and demand analysis.

One all-important reservation must be registered as this chapter is concluded. It concerns what may be termed unrealistic aspects of an approach such as has been presented here. A lack of realism in the formulation can conceivably stem from the specific behavior patterns within a particular industry which diverge from "expected" behavior and which result in a sequence of action other than what would occur if purely economic variables were in control. In such cases, the relative usefulness of the analytical framework falls, and therefore, it is mandatory to have this information.[93] In order to allow evaluation of the relative "realism" of the foregoing approach as applied to the steel industry, we must turn to a consideration of some aspects of entrepreneurial behavior in the industry.

93. A form of recognition of this problem is reflected in the following observation of Duesenberry and Grosse: "Finally we come to the problems raised by the variability of input coefficients among plants producing the same product by the same process. Studies of the chemical and electric power industries show that the capital input coefficients of plants producing the same product with the same material by the same method may vary from one another by as much as 50 per cent. A number of reasons such as special locational factors or managerial differences could be used to explain these differences. But at the level of detail ordinarily considered in input-output analysis or, indeed, in economics in general, these differences are not taken into account. They have to be regarded as introducing a random element into our coefficients. The significance of this element depends on the degree of correlation between changes in 'plant mix' from any base year and the deviations of the individual plant coefficients from their base year mean (calculated by weighting the coefficients of different plants by their base year outputs). Little can be said as to whether any systematic correlation exists, but even if it does not, there is bound to be a non-zero observed correlation resulting from the changes in plant mix from one year to another. Consequently, no matter how much we disaggregate or how carefully we measure, our coefficients contain a certain irreducible stochastic element. Whether we aggregate products or let the input-output system do it for us, a large part of this statistical error in our estimates of the sales of any industry will cancel out in the end. We can only judge by empirical evidence whether this source of variation in coefficients is important or not" ("Technological Change and Dynamic Models," pp. 82–83). However it should be noted that Duesenberry and Grosse have confined their need for "stability" to the immediate uses of input-output analysis, whereas this inquiry hopes to integrate demand formulations with macro-policy problems that will not permit reduction of managerial differences to a stochastic parameter.

FIVE FACTORS IN ENTREPRENEURIAL DECISIONS WITHIN THE INDUSTRY

In this chapter we will not attempt to present a complete profile of the steel entrepreneur. We will confine ourselves to factors which have bearing on the investment and capacity problem. This does, however, involve touching on a variety of factors at work in the over-all decision-making process, and we will be able to use evidence of attitudes and behavior in a wide range of situations. Note also that much can be found in the literature on management behavior, the relation of oligopoly to investment and innovation, and the effects of capital availability on expansion which has bearing on the steel case.[1] We will make explicit reference to some of this, but for the most part we are committed here to presenting empirical material about organizations, behavior, and decisions in the steel industry.

1. Cf. bibliography in M. Abramowitz, "Economics of Growth," and A. G. Papandreou, "Some Basic Problems in the Theory of the Firm" in *A Survey of Contemporary Economics, 2,* ed. B. F. Haley (Homewood, Illinois, Irwin, 1952); also F. and V. Lutz, *The Theory of Investment of the Firm* (Princeton, Princeton University Press, 1951); P. S. Florence, *Investment, Location, and Size of Plant* (Cambridge, Eng., Cambridge University Press, 1948); P.W.S. Andrews and E. Brunner, *Capital Development in Steel* (New York, Kelley, 1951); A. G. Hart, *Anticipations, Uncertainty, and Dynamic Planning* (New York, Kelley, 1951); G. Terborgh, *Dynamic Equipment Policy;* J. Dean, *Managerial Economics* (New York, Prentice-Hall, 1951); *Regularization of Business Investment, Report of the National Bureau of Economic Research* (Princeton, Princeton University Press, 1954), especially M. G. de Chazeau, "Regularization of Fixed Capital Investment by the Individual Firm," pp. 175 ff.; and, regarding the sources of funds to finance capital expansion, S. P. Dobrovolsky, *Corporate Income Retention, 1915–43* (New York, National Bureau of Economic Research, 1951).

The factors referred to may be placed in two general categories—namely, those aspects of decision-making which are a manifestation of the operating practices and the organizational structure of the individual firm and the market in which it is found, and, on the other hand, those which are conditioned by the general environment in which the entrepreneur works, that is, in the sense that historical development, technological change, and shifts in political and social factors may impinge upon the decision-making process. This chapter will deal with both of these categories.

ORGANIZATIONAL HISTORY AND MARKET BEHAVIOR

In Chapter 1 we suggested that a transformation has taken place in the organizational and institutional setting of business activity; what we find today differs from the pattern of an earlier era. The broad outlines of these changed organizational forms suggest that we may expect changes in procedure and behavior at all levels of operation, if a particular industry has indeed experienced such a transformation. Knauth describes this changing pattern:

> For well over a generation, businessmen without ultimate aim other than survival have, bit by bit, in adapting themselves to new conditions, pieced together this new form of economy. Managerial enterprise is a system of production and distribution, unified by policies and controlled by managers, whose main idea is to administer the business that concerns them in the interests of continuity.[2]

Knauth goes on to draw contrast between the "contours" of what he refers to as managerial enterprise as opposed to free enterprise. In managerial enterprise he sees the emergence of sunk capital, standardization and relative rigidity, concern for

2. O. Knauth, *Managerial Enterprise: Its Growth and Methods of Operation* (New York, Norton, 1948), p. 11.

the future, a separation of ownership and management, the ability of each unit to affect the market, market conditions shaped by policies, increased demand having its main impact upon production, cost based upon actuarial assumptions, and a goal of a stable and adjustable system, the functioning of which creates profits. He sees these conditions juxtaposed to the earlier free enterprise system of fluid capital, flexibility and mobility, concern for the moment, the identity of ownership and management, unimportance of the individual entrepreneur, action adjusted to impersonal market conditions, increased demand having an effect on price, costs estimated accurately, and a goal of immediate profits.[3]

After drawing attention to the concept of harmony of interests under the free enterprise system, he suggests that managerial enterprise "at times seems to pursue policies beneficial to itself but harmful to the community," though he points out that this will not necessarily happen. He summarizes his characterization of the new era as one in which the goal of management is an equilibrium situation. He goes on to say:

> Both growth and deterioration destroy equilibrium. Restoration is a continuous process. Relying upon its experience and its penetrating vision, management arranges and revises policies so that it can survey in the present and face the future with a favorable trade position. Frequently, one policy must be sacrificed for the sake of others. In reconciling conflicting policies, management is more concerned with its future trade position than its present. For today is tomorrow's yesterday.[4]

Knauth suggests an evolution of values which, in turn, suggests a change in the collective personality of the leadership group. In order to check this characterization against the experience of the industry, we will use a study of steel organizations over

3. Ibid., p. 33.
4. Ibid., p. 79.

the past fifty years by Schroeder[5] and a more detailed examination of the history of the United States Steel Corporation.[6]

UNITED STATES STEEL CORPORATION

In the decade following the turn of the century, two basically separate types of operation were to be found in steel. On the one hand, there was a dominant firm, the Steel Corporation. Its leader, Judge Gary, was preoccupied with the need to consolidate his already achieved power position, to fend off advances from competitors which would threaten his firm's position of leadership, and to justify and defend his actions and his accumulated concentration of power against a rising wave of public criticism. On the other hand, there were many other producers in the industry, none of which approached the position of the Steel Corporation, and all of whom were interested in the possibilities of growth and strength through merger and expansion of their facilities, against a background of relatively meager beginnings.

Carnivorous and belligerent behavior might have been expected of the Corporation; quite the contrary occurred. There was a conscious effort "to get on with one's fellows." Raid and plunder, the dumping and cut-throating actions familiar in other arenas of the period, did not appeal to Judge Gary, and so at the beginning the Corporation followed a pattern of behavior that set it apart from what might be termed typical of the times. Incidentally, this reflected the dominance of one man's personality and values over the affairs of the Corporation. This pattern of individual dominance, with its enduring imprint on the course and direction of policy, has been characteristic of the United States Steel Corporation through its history. It is a pattern facilitated from the very beginning, as

5. G. G. Schroeder, *The Growth of Major Steel Companies, 1900–1950* (Baltimore, Johns Hopkins Press, 1953).

6. "U.S. Steel," *Fortune, 13* (March, April, June, 1936); H. Maurer, *Great Enterprise: Growth and Behavior of the Big Corporation* (New York, Macmillan, 1955); E. B. Alderfer and H. E. Michl, *Economics of American Industry,* 3d ed. (New York, McGraw Hill, 1957).

early as 1901, by management control of proxies. The management, in effect, has been able to determine the membership of the Board of Directors, and with remarkable circularity the Board has controlled management. It is true, of course, that an area of concern to one or more members of a Board in a given period of the Corporation's history has differed from that of another period. For example, early concern with questions of markets and price stability gave way at a later period, under Benjamin Fairless, to union-management relations. At present, in the case of Roger Blough, there has been a reappraisal of industrial relations and productive efficiency in the organization.

The question of outside influence upon the affairs of the Corporation reminds us that at the beginning the influence of investment banking interests (e.g., the House of Morgan) was indeed strong. But in examining the contemporary situation we find that, though there are emphatic management assertions of preoccupation with finding sources of finance, the element of relative independence from sources of external financing, through an increased ability to draw on internally generated funds, places banking interests in a role far different from that which they held at the founding of the Corporation. In the '30s, the banking influence was still apparent, though there was even then little doubt as to the strength of the management of the Corporation to determine its course.[7] (It should be noted that in the structure of the Corporation, the President has always had a subsidiary position to the Chairman of the Board; this was, of course, true over the period 1901–27 when Judge Gary dominated U.S. Steel.)

Myron Taylor followed Judge Gary as Chairman in 1929, and was in turn followed in 1938 by Edward R. Stettinius, Jr. After two years in that position Stettinius was succeeded by

7. "The Corporation is a leading example of what is known as 'a Morgan company.' Indeed, considering its size and the fact that the elder Morgan created it, the Corporation may be called *the* Morgan company." Of the fifteen gentlemen on its Board of Directors six are Corporation men and six are 'Morgan men' (either partners or closely connected)." *Fortune, 13* (March, 1936), 63.

Irving Olds in 1940. The Board of Directors has typically consisted of fifteen men. Over the period 1901 to 1950, some 56 different individuals have served on the Board.[8] U.S. Steel's charter, in 1901, was the result of the merging of eight large companies which themselves were the result of previous mergers[9] and in 1901 it added four others.[10] Since that time, however, the principal additions[11] have been few and far between: Union Steel Company in 1903, Clairton Steel Company in 1904, Tennessee Coal, Iron and Railroad Company in 1907, and then none until 1930, when three additional firms were added, Columbia Steel Company in 1930, Atlas Portland Cement Company in 1930, and Oil Well Supply Company in 1930. The Consolidated Steel Company was added in 1948, the result of a post-war acquisition of plant created for war needs. The relatively few acquisitions of the property of competitors results directly from the early policy of Judge Gary—a policy of relatively slow growth and of "good behavior" in relation to competitors.

When Myron Taylor took the chairman's chair, he inherited not only a far-flung empire of semi-autonomous units, but much of the Gary policy of anxiety over public protest and a desire to disguise any appearance of monopoly that might come to the surface and hence to the public view.[12] As *Fortune* put it (in a contemporary view of the situation): "The dilemma

8. These and other figures on acquisitions and officers of the steel companies are from Schroeder, *Growth of Major Steel Companies,* pp. 36–78 and appendices.

9. American Tin Plant Company (1898) comprised of 36 companies, accounting for approximately 75 per cent of national tinplate capacity; American Steel and Wire (1898), 19 companies, 80 per cent of national capacity of wire and wire products; National Tube Company, 21 companies, approximately 80 per cent of wrought pipe and tube; American Steel Hoop, 9 companies; American Sheet Steel Company (1900), 26 companies; Carnegie Company (reorganized 1900), 19 companies; Federal Steel Company (1898), 5 companies; and National Steel Company (1899), 7 companies.

10. American Bridge Company (1900), 27 companies, controlling 50 per cent of structural fabricating capacity; Shelby Steel Tube Company, principal competitor of National Tube in the production of seamless steel tubing; Lake Superior Consolidated Iron Mines; and lastly the Bessemer Steamship Company.

11. Some 32 other, though not as significant, acquisitions did occur over that period; cf. Schroeder, *Growth of Major Steel Companies.*

12. During the period 1927–29, the Corporation was run by three men:

Myron Taylor faces, and has long faced, is how to give the subsidiaries enough autonomy for efficient operation and at the same time keep them 'corporation-minded' enough so that they put the general interest of the Corporation before the specific interests of their own province. But how efficiently a company so huge and with such diverse interests *can* be run by one man in one place remains as yet an unanswered question."[13] In the *Fortune* analysis of the problems of the Corporation in the mid-'30s, the point emphasized was that the previous orientation of the company had been one of delivering adequate returns on invested capital—"the banker's point of view" as opposed to the industrialist's—which was, as expressed there,

> how much had it been interested in protecting its investment (which means stabilizing) and how much in making and selling steel (which means pioneering)? . . . Any radical change in steel technology would render worthless hundreds of millions of dollars for the Corporation's plant investment. It is not surprising, therefore, to find that the Corporation had contributed very little to the art of making steel.[14]

Andrew Carnegie had observed cynically, "Pioneering don't pay." *Fortune* claims that this sentiment was echoed in a phrase once widely quoted as the Steel Corporation's policy: "No inventions, no innovations."[15]

Fortune describes the major problems faced by Taylor when

Farrell, as the Corporation's chief executive for operations, Myron Taylor, chairman of the Finance Committee, and J. P. Morgan, as chairman of the Board. Then, in 1929, Taylor took over.

13. *Fortune, 34* (March, 1936), 169.

14. Ibid., p. 172.

15. Ibid., pp. 173–74. "According to the report of a firm of industrial engineers engaged in 1936 by U.S. Steel to survey its properties, the Corporation lagged behind the rest of the industry in adopting a number of improvements in the production process: it was slow to adopt the continuous strip mill and the heat-treating process for the production of steel sheet; it was slow in getting into the production of stainless steel, cold rolled sheet, and other cold rolled steel products; it lagged behind in developing the uses of tinplate and exploiting the market for wire; and it was slow in utilizing waste gases from the furnaces and in utilizing low cost water transportation." Schroeder, *Growth of Major Steel Companies*, pp. 112–13.

he took over. They were: "One, a heavy burden of interest charges; two, inefficiency in making steel, which means chiefly too many old plants in uneconomic locations; three, inefficiency in selling steel, which means chiefly an archaic sales set-up and a lack of aggressive salesmanship; four, an unwieldly corporate organization with too much power at 71 Broadway and not enough in the operating companies; five, an inherited personnel and a consequent need for new blood."[16] According to their report, the action taken by Taylor in 1935 of merging the two largest subsidiaries, Carnegie Steel and Illinois Steel, and bringing in Benjamin Fairless from the outside to head this newly formed subsidiary revealed the Taylor policy: a deliberate attempt to give more autonomy to operating companies and take it away from the New York office. It announced Taylor's recognition that staffing was a very serious problem in the Corporation. The Board of Directors was made up of old men. Taylor set age limits of 65 for retirement and compulsory retirement at 70. He brought men in from other companies, notably Fairless and Edward Stettinius from General Motors. He hired a management consultant to survey the personnel situation, and was aware of the need to characterize the Corporation as a desirable place for actively interested steel men.

> From the very beginning the Corporation has suffered from a paucity of steel making and steel selling talent. . . . What was there to attract talent in the Corporation as Gary ran it—an absolute monarch who kept the purse strings entirely in his own hands? . . . Under such conditions it is hardly surprising that the management has never been particularly alert, imaginative, or aggressive.[17]

The Corporation was hard hit by the depression of the early '30s, and Taylor's recognition of its deteriorating position undoubtedly did much to stimulate action. Concentrating

16. *Fortune, 34* (1936), 180.
17. Ibid., pp. 186–88.

productive activity in the more efficient plants, rounding out facilities in the Pittsburgh area, and the consolidation already noted marked the first beginnings of Taylor's new pattern for the Corporation. Alderfer and Michl point out another reason why it made sense to decentralize managerial responsibility and concentrate more authority in Pittsburgh:

> Buyers of steel had installed during the period under discussion systems of inventory control and were ordering steel in smaller quantities and to more exact specifications. Such orders could be handled to advantage only by companies that had flexible organizations and policies.[18]

If the story were carried this far and no further, it would seem to suggest the desirability of decentralization; the experience of U.S. Steel up to the period of the '30s, and the relative success of Taylor in his actions to place more authority away from New York, appear to attest to this. However, as the history of the Corporation is traced further, we find, in the early '50s, an emerging awareness that the firm had reached a point of diffusion of responsibility that had now become a major source of trouble. With the succession of Fairless to the chairmanship of the Board succeeding Irving Olds, a new reorganization of the Corporation took place, reversing the moves of the '30s and indicating a tightening of the corporate structure. Taylor had sensed some of this problem at the end of his tenure. In 1938, he appointed a top management team, with Olds as Chairman of the Board, Enders as Chairman of the Finance Committee, and Fairless as President of the Corporation, but this triumvirate apparently did not have the dynamic effect that the Corporation still needed.

> There have been two major steps in the change-over. At the end of 1951 the triumvirate combined into one subsidiary the many that were engaged in the various phases of steel making. At the beginning of

18. Alderfer and Michl, *Economics of American Industry*, p. 82.

> 1953 it united this subsidiary directly with the Corporation. Thus the old company—formerly a loose aggregate of semi-independent units, each with its own board of directors—became integral divisions of a single organization. And thus that organization began to resemble its tight-knit profitable predecessor, the Carnegie Steel Company.[19]

This brings our narrative to 1955, when Roger Blough became Chairman, and to the contemporary organizational structure of the Corporation. Reference to some of Mr. Blough's attitudes will be made later in this chapter. A review of changes in the Corporation's structure up to this point can be usefully concluded with words of Blough's predecessor, Benjamin Fairless:

> But I do take great pride in having been a part of today's type of corporation management and perhaps in having helped to mold it a little. The trouble with the oldtime management was that it was just a little too individualistic. Practically every company started with some outstanding individual. He founded the company, built it and ran every aspect of it with an iron hand. When he died the company usually died with him, for none of the heirs could fill his shoes. In my time I have seen dozens of companies go downhill in this fashion, and often whole cities with them. Their downfall was sealed the day the founder died, or sometimes when he just got so old that his judgment was impaired.
>
> Somebody finally got the idea of hiring professional managers to replace the lone wolf genius who had bequeathed the company to his heirs. Thus was born the modern idea of corporate management, and thus was an opportunity given to people like me who were outsiders, often humbly born and not members of the clan. Modern management, with its profes-

19. Maurer, *Great Enterprise,* p. 248.

> sional attitudes, its concern for the public welfare and public relations and generally humane attitudes toward business, is far better than the old system. Some critics—*Fortune* magazine is one of them—deplore the fact that all businessmen today seem to be modest and polite duplicates of one another instead of flamboyant 'characters' as in the old days. But you cannot have it both ways, and today's 'man in the gray flannel suit' is ten times more efficient than his more colorful, more hot-tempered predecessor.[20]

BETHLEHEM STEEL COMPANY AND INLAND STEEL COMPANY

In the case of Bethlehem Steel, we find a different development pattern. As we have seen, the Chairman of the Board, particularly at the time of Judge Gary, dominated U.S. Steel. During the Gary period, for two brief years, Charles Schwab filled the office of President, but struggles with Gary were too much for him and he resigned to develop the foremost challenger to the Corporation's position, Bethlehem Steel. Schwab set the pattern for recruitment and organization of management in Bethlehem that survives today and marks Bethlehem somewhat apart from other companies. Bethlehem was founded in 1904 as a successor to the United States Shipbuilding Company, which had failed, and which owned the Bethlehem Steel Company and eight others. Bethlehem Steel dates back to 1857, and was the largest of the holdings of U.S. Shipbuilding. When Schwab assumed the presidency of Bethlehem, he took great pains to recruit a top management staff and establish a policy of promotion from within the ranks. This, combined with an executive bonus system and with the location of the top management people in Bethlehem, Pennsylvania—where they have easy and frequent access to each other for informal as well as formal

20. B. F. Fairless, *It Could Happen Only in the U.S.* (New York, Time, Inc., 1956), p. 37.

exchange of ideas—established the Bethlehem pattern. The company has had three presidents since 1904, Schwab, Eugene Grace, and Arthur B. Homer. Over the 50-year period of the Schroeder study, it should be noted that, excluding the men who were retired by Schwab in the first year, the company had only 14 different men fill the offices of its usual three vice presidents, secretary, and treasurer, and only 51 men on the Board. Bethlehem has a record of many acquisitions over the period 1909 to 1948 and, of course, recently embarked on a move to merge with Youngstown Sheet and Tube, blocked in the Federal courts by the Justice Department.

The third company to which some detailed attention will be given later in this chapter is Inland Steel. The unique aspect of Inland has been both its concentration of energies in one geographical location—Indiana Harbor, Indiana—and also the concentration of management over the years within one family stemming from the original founder in 1893, Joseph Block, and continuing to the present-day, when three members of the Block family have top management positions. Though it has been involved in merger attempts from time to time, essentially Inland has restricted itself to one operation and to activities directly within the steel industry. Where it has merged, notably with Ryerson in its desire to acquire warehouse facilities, it has maintained a scale of operations and a management attitude which concentrates on personal relations with its customers.

Though the record indicates that up to the '40s Inland had been the "most profitable" of the companies,[21] a summary table of annual rates of return in the industry will put the historical "pacing" as well as the absolute levels of return in better perspective. The data are presented in Table 18.

The relative importance of acquisitions in the expansion of U.S. Steel, Bethlehem, and Inland, compared with nine other

21. U.S. Congress, *Investments, Profits, and Rates of Return for Selected Industries, A Study Submitted by Federal Trade Commission to the Temporary National Economic Committee,* 76th Congress, 3d session, part 31 (Washington, Government Printing Office, 1941), p. 17618.

TABLE 18. *Rates of Return on Total Investment for the Principal Steel Companies, 1917–38*[a]

Year	*United States Steel*	*Bethlehem Steel*	*Republic Steel*	*Jones & Laughlin Steel*	*Youngstown Sheet & Tube*	*National Steel*[b]
	%	%	%	%	%	%
1917	30.94	20.89	40.74	33.69	55.62	
1918	24.60	9.52	18.75	13.55	18.09	
1919	9.28	11.00	4.34	14.90	9.40	
1920	10.65	7.52	13.83	18.49	10.26	
1921	4.39	6.13	6.75[e]	1.44[e]	.17[e]	
1922	4.65	3.69	2.04	3.87	5.13	
1923	9.32	5.69	11.32	7.78	11.26	
1924	7.55	3.88	4.26	6.20	6.83	
1925	7.84	5.07	6.89	6.85	10.23	
1926	9.23	6.05	8.56	9.74	11.03	
1927	7.25	4.90	5.82	7.10	5.49	
1928	8.71	5.48	7.29	9.26	7.69	
1929	12.18	8.93	11.52	11.56	12.92	
1930	6.16	4.71	.24	5.06	5.19	9.85
1931	.95	1.10	2.21[e]	.84	1.22[e]	5.78
1932	3.52[e]	2.03[e]	3.90[e]	3.89[e]	4.07[e]	2.83
1933	1.75[e]	.41[e]	.53[e]	2.24[e]	1.93[e]	3.85
1934	.81[e]	1.21	.08[e]	1.34[e]	.90	6.66
1935	.63	1.97	3.75	.06	3.12	10.33
1936	4.56	3.72	6.35	2.95	7.50	11.38
1937	8.64	6.92	5.65	3.47	8.49	15.44
1938	.22	1.97	.95[e]	1.79[e]	1.33	5.98
Annual average	7.33	4.53	3.84	6.03	6.45	8.17

TABLE 18—*Continued*

Year	*Inland Steel*	*American Rolling Mill*	*Wheeling Steel*	*Otis Steel*[c]	*Pittsburgh Steel*[c]	*Combined*
	%	%	%	%	%	%
1917	52.79	28.65	53.49		36.90	31.86
1918	26.05	42.50	27.40		25.49	21.95
1919	10.41	13.61	9.28	8.68	9.49	9.75
1920	11.86	19.89[d]	38.70	7.47	7.98	11.52
1921	.95[e]	8.00[e]	2.89[e]	22.20[e]	5.92	3.12
1922	.44	8.02	1.33	.02[e]	2.05	4.20
1923	10.95	10.82	7.57	5.90	7.63	8.68
1924	9.67	8.78	2.70	2.43[e]	6.08	6.45
1925	8.43	8.83	6.63	7.36	3.65	7.34
1926	10.40	11.98	7.62	8.67	8.55	8.77
1927	9.96	9.97	6.33	7.23	5.66	6.72
1928	13.88	13.96	8.78	13.12	3.87	8.27
1929	16.63	10.46	9.61	13.06	10.93	11.53
1930	8.95	2.37	4.05	4.61	4.90	5.47
1931	3.44	.81[e]	1.61[e]	2.44[e]	2.09[e]	.58
1932	1.39[e]	.24	2.80[e]	6.77[e]	3.86[e]	2.96[e]
1933	2.40	1.78	.77	2.79[e]	3.89[e]	1.03[e]
1934	6.73	4.04	2.13	4.92	1.84[e]	.41
1935	12.82	7.93	5.46	11.24	2.85[e]	2.42
1936	14.20	9.68	6.06	10.95	.50	5.52
1937	13.15	9.37	5.84	10.44	5.18	8.16
1938	5.37	.60[e]	1.99	1.60[e]	.57	.90
Annual average	10.18	6.53	6.09	4.18	4.92	6.59

[a]Source: U.S. Congress, *Investments, Profits, and Rates of Return for Selected Industries,* Study submitted by Federal Trade Commission to Temporary National Economic Committee, p. 17,621.

[b]Data are not available prior to 1930.

[c]Data are not available prior to 1919.

[d]Rate of return for 18 months; on a mathematical ratio it would be 13.26 per cent for 12 months.

[e]Denotes loss.

companies over the period 1901–50, is reflected in the Schroeder table (Table 19).[22]

TABLE 19. *Relative Importance of Acquisitions in Gross Fixed Assets Expansion of Major Steel Companies, 1900–50*[a] *(dollars in thousands)*

Company	*Period covered*	*Total additions to gross fixed assets*	*Total additions due to acquisitions*[b] *Amount*	*Per cent*
U.S. Steel	1901–1950	$5,071,743	$1,518,419	28.7
Bethlehem	1905–1950	1,629,794	460,300	28.6
Republic	1899–1950	702,143	319,262	50.0
Jones & Laughlin	1923–1950	467,381	53,137	11.4
National	1929–1950	344,784	109,654	31.7
Youngstown Sheet & Tube	1900–1950	464,044	134,850	29.1
Inland	1893–1950	366,655	15,275	4.2
Armco	1900–1950	349,766	64,014	18.3
Wheeling	1920–1950	264,559	68,347	25.8
Crucible	1900–1950	208,604	45,794	22.0
Pittsburgh	1900–1950	112,831	14,126	12.5
Sharon	1900–1950	63,304	16,561	26.2

[a]Source: Schroeder, p. 93.
[b]These figures include the value of fixed assets taken over at formation by those companies that originated from consolidations.

THE DECISION-MAKING FRAMEWORK

Decisions to expand capacity are of two primary types. There is expansion designed to take advantage of the opportunity to reduce costs through organizational rearrangement or utilization of new technology; there also is expansion which will result

22. Though questions of market structure and the implications for our discussion will be left to Chapter 6, we might quote Schroeder's conclusions on this subject here: "The statement may be made here . . . that the findings of this study of the growth of the twelve major steel companies suggest that the motives for integration may have been mainly 'political' rather than economic, and that the advantages, if any resulting from both forward and backward integration do not appear to have been primarily economic in the technical sense of the word" (*Growth of Major Steel Companies,* p. 128).

in additional revenue because of changes in the demand the firm faces, local or national. These may be said to be of one type.[23] On the other hand, motivation to expand capacity may be a matter of market strategy which stems from a desire to affect the relationship of the firm to its competitors, or, for strategic reasons, to affect the firm's place in the national economic picture.[24]

If one can generalize about the differences in the characteristics of these two types of expansionary moves, the latter, stemming from a desire to affect the "place" of a firm in the industry or in the economy, is more comprehensive in nature; strategic expansion is initiated after considerable introspective policy discussion, at high administrative levels and set against a planning horizon cast farther into the future. These are the types of decisions, for example, that lead a firm to include, for the first time, departments of basic research into new types of commodities. They also may express a firm's desire to affect its community standing, employee relationships, or intangibles such as the firm's goodwill. It is to be expected that this type

23. Among the theoretical observations which concern the steel industry are those dealing with whether the concept of profit maximization is as appropriate as "preference-function maximization" when speaking of the dominant rationale behind investment decisions. Cf. Papandreou, "Some Basic Problems," pp. 207 ff.; also K. Boulding, "Welfare Economics," in *A Survey of Contemporary Economics, 2,* 28.

24. A summary of findings of several studies of general management attitude and motivation, the nature and extent of forward estimates and capital expenditure plans, formal aspects of approval of capital expenditures and of factors determining the level of investment appears in R. Eisner, "Interview and Other Survey Techniques and the Study of Investment," in *Conference on Research in Income and Wealth* (New York, National Bureau of Economic Research, 1953), pp. 21–65; this study was later modified and published as *Determinants of Capital Expenditures: An Interview Study, Studies in Business Expectations and Planning* (Urbana, University of Illinois, 1956). Eisner cites a letter from a steel company president: "In my opinion, the necessity to bring costs down to 'a competitive basis' or lower if possible, is a strong influence on capital expenditures irrespective of whether the company is operating at capacity or below capacity. It may well be that greater attention is given to this problem in the latter periods when there is little necessity to expand, but prudence should dictate a continuing active interest in the subject. As far as our company is concerned, we are interested in spending capital to effect cost savings regardless of the situations, actions or prospective actions of competitors. Naturally, if a competitor develops a process which greatly lowers cost as compared with our own, it is a spur to action along similar or improved lines" ("Interview and Other Techniques," p. 57).

of planning is undertaken with less attention to capital budgeting and fine marginal calculations, and it often originates at the top. Commentary is requested from subordinates, but the source of initiative is with the senior executives or the directors of the firm.

The other type of expansion planning often originates from subordinates in various departments, and is submitted as part of departmental reports to the senior executives of the firm. Documentation for the cost-reducing or revenue-producing type of expansion is precise and very often is substantiated with short-term forecasts of market conditions or with the submission of engineering reports which verify the cost-saving potential of a particular program.

It will be part of our task in subsequent sections to establish what types of factors top management considers, in planning so-called strategic expansion, but we might note here that general attitudes and information on the state of demand in the economy and the total environmental picture will weigh heavily in expansion moves of the strategic type; attitudinal material provides relevant evidence in surveying this type of decision-making. On the other hand, techniques of short-term forecasting are the procedures used to formulate expansion plans of the cost-reducing, revenue-affecting type, and we need to examine other evidence to shed light on them. Before moving into specific attitudinal examples, we will look at some of the problems that accompany planning for expansion in more general terms, i.e., questions on the types of equipment, materials, surveys of the market, and methods of production.

OPERATIONAL PRACTICE, TIMING, AND PLANNING HORIZONS WITHIN THE FIRM

A limitation on the factors that might be covered under the above heading is necessary, because our emphasis is so specifically focused on the timing of investment decisions within the industry. This means that a considerable range of factors will be treated in less detail, i.e. those which generally qualify as determinants of the level of expenditures. For example, in addition to such items as who proposes and who approves expenditures, they include the basis on which particular projects are approved or rejected, matters of costs,[25] availability of funds, turnover period,[26] minimal return,[27] and so forth.

These will be slighted in favor of emphasis on questions of timing.[28] This is justified on the grounds that the focus through-

25. A considerable number of studies have been devoted to cost conditions in the industry (cf. note 22, in Chapter 6), some of which relate capacity utilization to cost consideration. For example: "In the case of the United States Steel Corporation, fluctuation in the percentage of capacity operated was the most important single factor associated with variations in cost per ton. Fluctuations in wage rates in general were accompanied by less than proportional variations in cost per ton or even in labor costs per ton . . . these conclusions are tentative." K. H. Wiley and M. Ezeckiel, "The Cost Curve for Steel Production," *Journal of Political Economy, 48* (1940), 813.

26. In such calculations there is indication of consistency with the accelerator relation. Leontief notes: "In general stock-flow ratios are more readily observable than lags between flows. It is interesting to note in this connection that the conventional standards of behavior, the rules of thumb actually, or at least apparently, adhered to by the practical decision-makers in economic enterprises—be they industrial managers, directors of commercial or financial institutions, or even public budget-making authorities—most often are formulated in terms of some normal period of turnover, desirable inventory ratios and other, similar stock-flow relationships; these conventional rules hardly ever contain explicit references to desirable or normal time lags" (*Studies in the Structure of the American Economy*, pp. 54–55).

27. Return, of course, involves pricing policy of firms in the industry. A recent comprehensive study of pricing treats in some detail pricing policy in the steel industry, specifically with reference to the United States Steel Corporation and National Steel. In this study, considerable attention is paid to target return pricing where the primary determinant is the maintenance of stable return as opposed to a simple profit-maximization motive. The authors point out that "a company that is highly conscious of its targets of return may in one case have a bold and promotional approach to price making, while another thinks of target return in terms of a cost-based structure of prices in which price stability has high priority. A company may, on the other hand, give a high degree of autonomy to product groups, which respectively exemplify both approaches, and have its co-ordination almost exclusively at the financial investment level without regard to the differences in price methods which enroll particular divisions to achieve profitable results." A. D. H. Kaplan, J. B. Dirlam, R. F. Lanzillotti, *Pricing and Big Business: A Case Approach* (Washington, The Brookings Institution, 1958), p. 8. Cf. also F. Machlup, *The Basing Point System;* Alderfer and Michl, *Economics of American Industry*, p. 85; C. Kaysen, "Basing Point Pricing and Public Policy," *The Quarterly Journal of Economics, 42* (1949), 289–317; and A. Smithies, "Aspects of the Basing-Point System," *American Economic Review, 32* (1942), 705–26.

28. Operation practice associated with some of these factors is reflected in Eisner, *Determinants of Capital Expenditures,* pp. 88–89.

out this study has been on the critical nature of capacity adequacy at certain points over the cycle, with a view to avoiding the potential bottleneck condition that might arise. For this reason, in conversations with members of the industry more attention was given to the time sequence in decision-making than to other aspects of the investment process.

Some of the judgments which are to follow will be based on what may well be considered an inadequate sample. However, for purposes of characterizing a possible set of relationships which may be checked against specific data, they seem to have genuine usefulness. This is particularly so because they cast some light on the accelerator relation and the relative likelihood of its presence within the decision-making framework. It should be noted here, however, that whereas in previous sections discussion was concerned with the possible effect of changes in output in the economy on the demand for steel, reference here is to the response of steel entrepreneurs to changes in demand for steel. Specifically, it would be desirable to know whether decisions to expand steel capacity occur *in the midst of* a change in the demand for steel or rather *in anticipation of* the change in demand. The former would suggest a closer tie to the pure accelerator, whereas the latter might indicate that expectations, regarding what demand might be some time in the future, control possible plant expansion planning and other investment.[29]

There are two elements which should be noted in connection with investment decisions which are implicit in the above, but which should be specified in our discussion. The first is the distinction between capital-saving and labor-saving technology. The incorporation of new methods of production which have their primary effect in their labor-saving aspect, of course have relevance for their impact on distribution of income in the economy, for questions of adequacy of labor supply, and demand for loanable funds as regards capital-saving technology. These are questions which involve the aggregate economy and

29. Cf. P. G. Clark, "The Telephone Industry: A Study in Private Investment," in *Studies in the Structure of the American Economy*, pp. 244 ff.

are distinct from the context of our present discussion, i.e., we are now thinking in terms of anticipation problems and planning problems within the firm. Only insofar as a firm is pressed to move toward capital-saving technology or toward labor-saving technology, when new methods of production appear because of contingent shortages in the respective factor markets, are the decisions to invest based, from the perspective of the firm, on a set of criteria which tends to merge these separate compartments.[30]

The other aspect of the investment question which is relevant when looking from within the firm to outside conditions is the role leadership-followership plays in adoption of new technology. Even in highly oligopolistic markets, the introduction of a radically new approach to production will compel the non-innovators in the industry to take heed and to modify their processes, so as not to be at a competitive disadvantage. Note, however, that even in terms of the competitive needs of a situation, the demand picture holds sway: "the reaction of the new product or method of production upon investment outlays is via the deleterious effect upon sales expectations. And substantial investment may be necessary to prevent sales and profits from declining; no actual increase in profits above current levels may be anticipated at all."[31] Buchanan quotes Charles R. Hook, president of the American Rolling Mill Company, before the TNEC:

> When a fellow comes along with a method of doing a thing that you have got to adopt to keep up with the pace, you go out and break your back to find the money to do that job, to keep from going out of business, and that is just exactly what happened. We have proved by the introduction of this process that a quality of material could be produced that could not be produced by any other method, and if they

30. Cf. W. Fellner, "The Technological Argument of the Stagnation Thesis," *The Quarterly Journal of Economics, 55* (1941), 638–51.

31. N. S. Buchanan, "Anticipations and Industrial Investment Decisions," *American Economic Review, 32,* supplement (1942), 149–50.

> wanted to keep in competition with the game, it was necessary to go out and put in one of these plants.[32]

This discussion may be opened by referring to the means of analyzing demand employed by a particular firm. From some statements of industry leaders in the post-World War II period, it would appear that demand is seldom analyzed in the aggregate or projected very far into the future. Those estimates that were made for five or ten years generally tended to gauge the trend of steel consumption on a per capita basis.[33] In the past, they seem to have served primarily to reassure the decision-makers that long-run prospects do allow a certain amount of leeway in planning for more immediate swings in demand. At the same time, the longer-run projections appear to have served as a planning ceiling, tending to discourage expansion to satisfy unusual demand of the moment, which management believed would leave the organization over-extended in a later period.

In contrast to various practices concerning demand analysis for the longer run, firms in the industry have based (and now base) estimates of more immediate demand on patterns of current and expected production of major consumers of the respective firm's output. If, for example, the firm delivers sizable amounts within a particular region or to a particular consuming industry, then, logically, preoccupation is with activity in one of the respective categories. In the case of one firm, for example, planning for future increases in output might be based on knowledge of specific additions to productive facilities in the automobile industry in the local area and on the assumption that a certain share of this market would be theirs. The potentially hazardous aspect of this type of dependency is seen in what considerable impact a loss, e.g. through fire (such as occurred at one of the General Motors plants several years ago), may have. Where a particular firm concerns itself with regional demand situations, because historical development of the firm has been such that the major portion of output has

32. TNEC, Hearings, part 30, quoted in Buchanan, "Anticipations," p. 149.

33. See, for example, earlier reference to W. Sykes, pp. 107–08.

been consumed in the region, questions concerning demand are centered on relative rates of growth in the region. It may be noted parenthetically that the analytical formulation suggested in Chapter 4 lends itself to this type of breakdown also.[34]

In surveying the approaches to forecasting used by the various corporations, it is hard to separate their short-term forecasting procedures from their attempts to project and plan for the longer run. This is true because in some of the procedures used there is a merging, and in some cases confusion, of short- and long-run appraisals and also because, of necessity, forecasts for purposes of planning plant expansion must involve a look into the relatively long-term future, in as capital intense an industry as steel. Indivisibilities of productive equipment and relatively long pay-off periods demand that even anticipation of short-run changes in demand push the executives to consider questions of the longer-run future before decisions to meet short-run expected rises in sales can be decided upon. But it should be noted that in more recent pronouncements, a change in perspective can be seen regarding longer-run projection and the use of macro-economic data. Management not only acknowledges concern with immediate and local conditions of demand, but asks questions about national economic activity. For example:

> Bethlehem management must keep an eye on shifts and changes in product market at all times. Our capital Statistical Research Division continuously makes available figures covering the past histories of all our products—sales totals, location of major purchasers, etc., for both our company and the industry—and we must then look ahead five, ten, even twenty years and estimate what and where demand will be for those products. Factors such as population growth,

34. Cf. W. Leontief and W. Isard, "The Extension of Input-Output Techniques to Interregional Analysis," *Studies in the Structure of the American Economy*, part 2; also J. H. Cumberland, "Locational Structure of East Coast Steel" (unpublished dissertation, Dept. of Economics, Harvard University, 1951).

> economic conditions, imports, competitive substitutes, technological advances, must be taken into consideration before future tonnage requirements for any product can be estimated.[35]

It is interesting to note that a firm, which ten years earlier had been noted for its outspoken denial of the "Bean outlook" as regards national need for steel, devotes in a later annual report a section to the "vastly expanded steel markets" which lie ahead in the next decade. In the 1957 annual report of Republic Steel, there is a section announcing the publication of a study just completed by Republic Steel, heralding the future "keynoted by growth." The previous spring a report, "U.S.A. Tomorrow," was published, in which Republic engaged in a ten-year projection of the American economy.[36] As conservative planners, the authors stress: "U.S.A. Tomorrow is neither a prediction nor a goal, but rather a conservative yardstick to measure the potential of the years that lie ahead." Yet their forecasts prescribe the growth that is to take place in national industrial capacity and G.N.P., and also the growth to be expected in science, research, the employment of a wide variety of materials, and particularly the expansion of the use of steel.

After referring to a population target of 190 million for 1965, they forecast a G.N.P. of $560 billion, a rise in over-all productivity of 28 to 30 per cent, and a per worker investment in capital equipment of $18,000. After forecasting capital outlays for 1965 at a $50 billion annual figure, a rise of $15 billion per year over 1955, they speak in terms of an annual ingot capacity of 160 million tons for steel, up 20 million over the 1955 figure. They then go on to speak in terms of increases in the steel consuming industries, of continual growth, particularly automotive, oil, construction, machinery, electric power, and other producers durables. It is interesting for our purposes to see a steel firm engaged in detailed projections of various

35. *Bethlehem Review, 75* (Bethlehem Steel Company, Inc., Bethlehem, Pa., January, 1959), p. 7.

36. Cf. *Annual Report, 1957* (Republic Steel Corporation, Cleveland, Ohio), pp. 14–15.

industrial sectors in the economy in order to provide a long-run base upon which to calculate its own needs for new capacity. But in the particular case of Republic Steel, there is in this report a clue to the modification of entrepreneurial attitudes in the industry that has taken place over the intervening ten year period. Republic's report states:

> The potentialities set forth in "U.S.A. Tomorrow" may very well prove conservative. Automation is moving ahead and at an accelerated pace, and another decade should see it revolutionizing many segments of American industry. Sharply increased expenditures on research and development could bring us a flood of products currently unknown or unimagined. Thus, if we make wise use of our technological, financial, and manpower resources, the future may far surpass today's view of tomorrow's horizon.

This sketches, in somewhat general terms, a type of planning projection. It might be useful to look in some greater detail at the planning procedures of a firm in the industry, and to this end we will draw on statements by staff members of United States Steel.[37]

In pointing up the interdependency of general indicators with items of specific relevance to steel forecasting, Rich's report includes a table of "Classification of General Business Indicators" (Table 20). It places those items of "background and intermediate information" to the left of general business indicators, and those which are "determinants of the final demand for steel products" to the right of general indicators. The items in the fifth column relate to specific steel demand; those in the fourth column impinge directly on final demand for the industry, but are more general in nature. The report indicates that department store sales and mail order sales are of con-

37. Particularly a report by J. L. Rich, United States Steel Corporation, "Techniques and Uses of Forecasts of General Business Conditions in U.S. Steel," *Proceedings of the Business and Economics Section, American Statistical Association* (Washington, 1956), pp. 161–64.

TABLE 20. *Classification of General Business Indicators*[a]

Background factors	*Intermediate*	*General indicators*	*Intermediate*	*Specific factors*
Internat. situation	Unemployment	Gross national product	Personal cons. exp.	Passenger cars
		Pers. cons. exp.		
Domestic conditions	Employment	Gross pvt. dom. inv.	Durables	Appliances
	Agricultural	Govt. purchases		Furniture
Population	Nonagric.		Nondurables	Metal cans
		National income		
Households	Productivity		Services	
		Personal income		
Labor force	Av. hrs./week	Personal taxes	Residential constr.	Pvt. construction
		Disposable pers. inc.	Nonres. constr.	
Armed forces	Av. earngs./hr.	Savings		Oil & gas drilling
			Expenditures for	
Civil. labor force		Corporate profits	prod. durables	Machinery
				Transportatn. eqp.
Raw material		Government receipts	Inventory change	
				Inventories
Energy supply		Manufacturers:	Defense expenditure	
		Sales (shipments)	Gov't. construction	Gov't. construction
Transport. supply		New orders	Other gov't. exp.	Aircraft
		Order backlog		Ordinance

TABLE 20—*Continued*

Background factors	*Intermediate*	*General indicators*	*Intermediate*	*Specific factors*
Education		Inventories	Farm prices	Shipbuilding
		Production	Farm income	
				Agricultural eqp.
		Wholesale and retail	Consumer intentions	
		Sales		Exports
		Inventories	Business spending plans	
		Dept. store sales		
		Mail order sales		
		New pass. car sales		
		Steel production		
		Freight car loadings		
		Electric power prod.		
		Wholesale prices		
		Consumer prices		
		Interest rates		
		Money supply		
		Bank reserves		
		Consumer credit		

[a]Source: Rich, "Techniques and Uses of Forecasts," p. 162.

siderable use to the Corporation as a specific series of high predictability. New passenger car sales are singled out as the "single most valuable measure of consumer well-being." "Other general indicators, often used in the development of weekly indexes of business activity, include steel production, freight car loadings, electric power production, wholesale prices and consumer prices."[38] Rich points out that the firm is becoming aware of the necessary parts that general business indicators play in forecasting for the industry or a particular commodity within the industry.

Data on population, number of households, labor force, raw material, energy supply, and transportation facilities all are particularly useful in longer-run forecasts. Rich points out that forecasts which draw only on projections of specific demand in consumer industries are likely to fall short of the demand that materializes in the long run. He cites limits to "human imagination" as a reason for this happening. When projections are based on more aggregative data, and past trends are used as a basis for projection, the whole complex of factors that make for growth in the economy is built into the projection, and yields higher estimates of future demand than will a more restricted projection of specific commodity markets. The other advantage cited for using general indicators in connection with more specific sector projections is that this tends to modify the effects of the natural imbalances that occur in the economy when one sector or another is ahead or behind; hence, the risks of being led astray by this likely imbalance are tempered. Rich is not unaware, however, of the limitations of aggregative data in terms of specific industry need (as we stressed in Chapter 4 above) and he points out specifically that

> a trend line through steel production, or a per capita projection, or a projection based on Gross National Product in total may give a useful guide to over-all steel demand, but yield no information whatsoever

38. Ibid., p. 162.

> with respect to product mix or geographical location of the new facilities. For this sort of detail, there is no substitute for a careful industry-by-industry forecast.[39]

The other half of the coin is the problem of forecasting in specific industrial markets.[40] The distinction that the forecasters in this area are cautioned to keep in mind is that of the sales versus the market forecast. The recommended procedure is for an analysis of the expected demand in a given market to be made and, following this, a sales policy designed to reach a certain degree of penetration in the predicted market. This is cited as "planning for sales" as opposed to "forecasting sales," whereas the anticipation of demand for product is considered an explicit forecast, obtained by means of a specific analytical framework. The value of the distinction rests with the fact that ex post analysis of accomplished sales can be evaluated for failure of sales to penetrate a correctly forecast market or, on the contrary, for a failure in results to be pinned to an inaccurate anticipation of market volume. There are advantages in forecasting both normative values and deviations around normative trends because, as far as planning in the industry is concerned, plant capacity may well have to be geared to esti-

39. Ibid., p. 164. Rich points out that G.N.P. projections used alone could be quite hazardous, because the type of final demand in the economy that makes up a given G.N.P. figure is crucial with respect to steel demand. He states in this connection: "It can be roughly estimated that a million dollars of personal consumption expenditures for durables is worth ten times as much as a million dollars spent on consumer non-durables in its effect upon steel demand. On the other hand, expenditures for new private construction are worth one and a half times as much as spending for consumer durables and expenditures for producers durables such as machinery, transportation equipment and agricultural equipment are worth two and one-half times as much as personal spending for durables. With differences like these, the pitfalls of matching against total G.N.P. are quite apparent."

40. Some of these illustrations are drawn from material presented by B. E. Estes of the United States Steel Corporation. Some references to forecasting procedures used by various firms will be based on examination of actual reports submitted to senior management by officials of the respective firms but will not be specified by the writer because of requests by management to hold the contents confidential.

mates of peak demand whereas profit expectations would be based on the mean trend.[41]

Questions of demand conditions and future expectation as well as present and future expected construction costs are as pressing as specific earning calculations. Andrews and Brunner point out that when capital outlays are made in anticipation of producing new products or using a new productive process,

> The decisions are so much a matter of broad views of trends of demand and of profitabilities broadly estimated, together with prestige factors sometimes involved in entering new markets that it woud be misleading to think of them as at all based on estimated earnings in a sense that critical minima are required before adopting a scheme. Everything is too conjectural for such rigid theoretical notions to be very helpful.[42]

Long-range forecasts, i.e., those covering from a ten- to a twenty-five year period, are made on an annual or bi-annual basis in some of the large steel firms. The procedures used in developing forecasts are fairly well established. An example of such an approach follows. The first step is to obtain a picture of the aggregate economic climate, which encompasses such questions as international relations, levels of employment, range of government activity, and assumptions about the rate of technological change, productivity, and questions concerning the labor force. The second step is to estimate, in quantitative terms, some of the aggregates for the economy. This is similar to the Rich approach discussed above. From estimates of overall economic levels, certain projections of steel demand are drawn. This involves a linking of the aggregate output of the industry to, say, levels of G.N.P. Next is a move to pin down

41. One-year forecasts are looked upon as holding great promise of accuracy; two- to five-year forecasts are looked on with reservations. Though acknowledgment is made for the need of this type of forecasting, the necessity of frequent revision is stressed.

42. Andrews and Brunner, *Capital Development in Steel,* p. 356.

demand in terms of product mix, through projection of steel consumers' markets and anticipation of their levels of production. This is then translated in terms of product mix for the steel industry itself. Ultimately, each firm makes decisions concerning the market penetration it will strive for so that its product mix will satisfy the particular demand it hopes to satisfy.[43]

From this, the logical step is to see whether facilities in existence and under construction will be adequate to meet anticipated demand. It is here that high level policy decisions enter forcefully. Decisions as to further penetration into markets already entered, and decisions to alter product mix for the firm, in anticipation of shifts in over-all components of demand, are made in executive session. The extent to which a given firm's administrators will call for forecasting studies such as those described above, and, once they have been delivered, the extent to which they will be drawn upon (in fact, even believed) will, of course, vary from firm to firm and will often serve as a basis for differentiating the type of leadership of one firm from that of another.

The research activity described here represents the outer limit of activity taken in the industry, as it describes the type of analysis undertaken by some of the larger corporations in the industry. It can be presumed that, with slight modifications of technique, the forecasting activities of United States Steel are as complex as any that exist within the industry. However, apropos of the discussion in Chapter 4, the present description also reflects the relatively simplistic methods even the Steel Corporation uses in anticipating demand. The most sophisticated techniques which have been developed in this area, consistent with the format described above, are probably those of Boschan (discussed in Chapter 4), but the development

43. Demand figures derived from an analysis of industry-by-industry consumption of steel should be consistent with analysis of aggregate steel demand derived from projections of per capita consumption of steel in the over-all economy. It is noted, however, that as projections move further and further into the future, industry-by-industry estimates of steel consumption, as opposed to the over-all linkage of steel to G.N.P., involve greater discrepancies. This, of course, is explained by the historic inability of the industry-by-industry linkage to anticipate new uses for the product.

of more elaborate characterizations of interindustry demand seems to lie in the future, as far as firms in the industry are concerned.

Some firms which, although they may have a market analysis department, do not have an elaborate statistical research division may rely on outside sources for their projections of the aggregate economy. It is not untypical for their own staff to take the aggregate projections of markets perhaps even broken down in geographical terms thus applied, and relate them to the particular penetration goals and product mix objectives of the firm. When a firm is particularly interested in the market trends of specific regions or in a specific product line, it must carve out of the aggregate data, trend data for particular regions or particular consuming industries.

Take the case of regional concentration: projections for population figures by region would be relevant, and projection might well involve something like a ten-year period. For this firm, it would not be the given percentage increase in population anticipated in the United States as a whole that would be used, but the fraction of this percentage which would occur in the firm's area. In this case, figures on the region's share of total value added for manufacturing in the economy would be reported; so would the share in production workers as a percentage of total U.S. employment. Other data would tie the region's steel-producing capacity to the particular district's share of national capacity and national production; to this would be added figures on the particular firm's share of industry production and capacity. A report on these trends might well include statements concerning the stability of growth of the region's share of national capacity and production, and also of the firm's relationship to these other figures over time. And information relevant to future possibilities as to continuation of trend might be qualified by recognized changes in fundamental conditions, e.g., the opening of the St. Lawrence Seaway and its impact on midwestern production. A firm, if it is focusing attention on the regional situation, is interested in establishing the net inflow or outflow of product to consumers in the area in question; hence, figures on production by the region or even by states within the region, and also on the amounts of various steel products consumed, are relevant data. Determining that the area is a net importer of steel will condition certain decisions on capacity planning.

A fundamental decision upon which the readers (the ultimate decision-makers) of these statistics must agree is whether they will assume a per capita increase in consumption of steel or whether they will base their planning on some fixed per capita amount. This judgment obviously affects the dynamics of the planning schedule; estimates based on fixed per capita production over time would lead to much more restrictive capacity planning. The point to be stressed here, however, is that reports to the directors present the figures in terms which leave this type of decision up to them, i.e., figures are typically presented in a way that leaves the decision-makers a range of alternatives where they make not only a judgment about the persistence of per capita trends, but also decisions as to whether the firm wishes to maintain, increase, or relinquish its current share of the market, either regional or national.

The research group presenting this set of figures typically would leave to the directors the question of whether to accept, for example, that the area would continue as a net deficit area. It also would be careful to present a range of alternatives, leaving to the directors the choice as to which levels of production and what product line the firm would preserve. This suggests that a very loose array of data is presented (undoubtedly partly because of the limitations on forecasting tools).[44] Estimated demand is only one of the relevant variables and, of course, decisions to expand capacity over the long run, though drawing on this material, would also be based on expectations as to cost, "ambitions" regarding market shares, and the range

44. Research undertakings to forecast demand for a particular product, for example sheet and strip, would follow a pattern similar to that above. Comparisons would be made between the firm's shipments and that of other producers in the area, data would be developed projecting demand by present consumers in the area and then a reconciliation of projected demand and firm share of the market would be presented.

of intuitions that are brought to bear on such decisions.

The question of the pitfalls and limitations as well as the strengths of forecasting is beyond our discussion. However, there is a point that might be made here.[45] Reference to the work of so-called sales and research divisions of the firms do give some sense of the raw data that are presented to the executives, who then proceed to supply the component of expertise in the decision-making process. One must keep in mind the fact that the use to which this kind of "professional" demand analysis is put will vary from firm to firm, and will depend on the firm's readiness to accept this kind of assistance. It also will depend upon the length of the planning horizon and the over-all scope of the undertaking under discussion. A firm's experience with the use of this kind of data tends to increase its readiness to make use of and strengthen departments engaged in this work. It is not untypical for a firm which in the past has relied on outside agencies (either private economic analysts or drawing on government data) to move in the direction of starting departments within its own organization and subsequently bolstering them, as acceptance of this type of analysis grows, and as the skills of statisticians in the field increase. Executives vary widely in their opinions of the helpfulness of this data; nevertheless, there seems to be a growing acceptance and understanding of the role this type of analysis can play in planning.

The impact of the firms' use of forecasting on aggregate forecasts of the industry's plans and expectations can be seen in the sharpening of the predictive skills of industry generally throughout the economy, though it has been observed in the McGraw-Hill surveys that large firms tend to overpredict and hence can throw aggregate predictions for the economy off. This raises the

45. For a discussion of some of the problems in forecasting, see, for example, A. M. Okun, "The Value of Anticipations Data in Forecasting National Product," in *Quality and Economic Significance of Anticipations Data,* National Bureau of Economic Research, Special Conference Series, 10 (Princeton, Princeton University Press, 1960); see also National Bureau volume, *Short Term Economic Forecasting* (Princeton, Princeton University Press, 1955).

question of self-fulfilling prophecies. As time passes and more reliance is placed on aggregate projections delivered by national data collecting agencies (both private and public), the over-prediction which is built into some of these estimates may lead to upward revisions on the part of individual firms drawing on the data. Hence, feedback may result in rates of expansion that would not have occurred in the absence of this upward bias in the original input of data to these national collecting agencies.

Another aspect of the forecasting question which must be kept in mind: a number of forecasting entities working apart from each other, though partially drawing on the same set of basic data, may be expected to come up with somewhat different forecasts. Although within each firm there may well be consistent use of data delivered by the respective research departments, action from firm to firm may vary, as the techniques, skills, and refinements of the forecasting procedure vary from firm to firm. A superficial manifestation of this can be seen in the following news release:

> A slightly larger 1958 industry output was predicted by a U.S. Steel executive last week. Marcus J. Aurelius, administrative vice president, predicted output in the vicinity of 90 million tons. His projection assumes the 'inventory correction' will have spent itself by the end of the third quarter. A. B. Homer, president of Bethlehem Steel Corp., said he anticipated improved steel demand in the third quarter, while Charles M. White, Republic Steel Corp., chairman, said that inventory liquidation should run its course not too long after mid-year. George M. Humphrey, chairman of the National Steel Corp. and former U.S. Secretary of the Treasury, told his company's annual meeting the third quarter is 'anybody's guess,' with vacation-time slowness possibly slowing the inventory reduction, but he predicted greater volume after Labor Day.[46]

46. *Wall Street Journal* (April 28, 1958), p. 26.

This would suggest the need for standardization of procedures in addition to more uniform access to relevant basic data.

Generalized remarks made in discussion with executives concerning their medium- to long-run outlook on demand disclose an apparent lack of clarity regarding the time dimension of demand analysis and planning at the intermediate level. At one end of the spectrum, long-run plans are discussed as part of general policy and programming in the firm, and may well involve periods of over ten years.[47] At the other is a framework for demand analysis and subsequent expansion planning within the one- to four-year interval as discussed above. Often plans for the purchase of a new crane or the addition of new facilities for handling freight are calculated as to cost, pay-off period, delivery date, and so forth.[48] However, the functional relationship between particular additions to capacity as related to anticipated increases in demand is undefined.[49] No unified sense

47. Eisner quotes the vice president of a large steel company: "We don't have long-run forecasts, we do have long-run construction plans.—How far ahead? Further than we live! . . . These plans run thirty or fifty years ahead. They are not fixed, of course. But when we fill in a little part we can have a picture of the whole. . . . I don't have too much faith in long-run forecasts. Things depend on national and international developments." Eisner, *Determinants of Capital Expenditures,* p. 53.

48. For discussions of procedures on capital budgeting and the specifics of investment programming, cf. Terborgh, *Dynamic Equipment Policy;* Dean, *Capital Budgeting and Managerial Economics;* and W. Heller, "The Anatomy of Investment Decisions," *Harvard Business Review, 29* (1951), 95–103.

49. Once again the difficulties in drawing the line between replacement and net capital formation should be emphasized. As with distinguishing between cost reduction and demand-induced incentives to alter capacity, this distinction is subtle and elusive. In fact, the idea of replacement investment itself has various shadings and consequently differing significance depending upon the particular outlay involved. Mrs. Mack gives us an example of this: "Viewed from the standpoint of similarity of equipment, it is clear that the continuous mill bears only a faint family resemblance to the one-stand 'hand' mill. On the other hand, the one does often physically replace the other. Although perhaps as many as ten hand mills may be dismantled and one continuous mill installed, viewed from the standpoint of similarity of product, it is clear that both mechanical and hand mills rolled sheet and plate steel. The mechanical mills (supplemented by cold-reduction finishing) have improved the surface, the uniformity of gauges (and the ability to roll narrow gauges in wide strips), and have resulted in a considerable reduction in price. Both improved quality and lowered price have tended to develop new uses for the output of the mill. They have also tended to expand sales in the old markets. (In part, also, the development of the mills was the effect rather than

of stock-flow relationships seems to exist (in discussion), yet implicit recognition is evident, for example, in references to plans to increase certain finishing capacity because of expected increases in orders from certain consumers.

Demand analysis, as indicated above, then appears to be in the hands of "experienced" sales representatives, who know the market and report back on likely prospects of increased demand as well as "trends" among consumers. And as regards a framework for demand analysis, planning practice appears to be at times on an ad hoc basis and at other times is expressed in terms of an over-all long range program.[50]

As regards timing of investment expenditures in relation to changes in demand levels, evidence is not clear cut. This stems in part from the diversity of planning practices suggested above, and also from the necessary consideration of those factors which distort characterization of the "pure" capital-output relationship. For example, anticipation of technological improvement may result in the forestalling of additions to blast furnace capacity. Another example is found in the Inland Steel situation, where excesses of finishing capacity have existed, and where we would expect to find reluctance to expand finishing capacity to a point where discrepancies between ingot production and finishing capacity would be exaggerated. As a matter of fact, in some cases Inland has expected consumers to provide raw ingots to be put through their mill, because of their relative deficit position regarding ingot production.

Figures for the output of Inland Steel, for example, when compared with net investment over the period 1933–52, do not yield any conclusive picture as to possible lag intervals between the series. At least, there is no consistent pattern of lag. These figures, shown in Table 21, indicate an output lead over investment of one year in downturns in 1937 and 1942,

---

cause of certain market situations.) If, finally, we relate old to new mill on the basis of amount of output, we must consider the situation at various levels." R. P. Mack, *The Flow of Business Funds and Consumer Purchasing Power* (New York, Columbia University Press, 1941), pp. 93–95.

50. Cf. Eisner, *Determinants of Capital Expenditures,* p. 83.

and of two years in 1946 (the years 1948 and 1951 coincide on turning points). In upward points, output leads investment by a two-year interval in the years 1940, 1944, and 1947 and

TABLE 21. *Inland Steel Company Production and Net Additions to Property,*[a] *1933–52*

Year	*Production (thousands of net tons)*	*Net additions to property*[b] *(millions of dollars)*
1933	946.6	4.9
1934	1,191.1	(1.7)[c]
1935	1,629.4	12.1[d]
1936	2,164.6	13.4[d]
1937	1,962.7	21.2
1938	1,416.3	6.3
1939	2,408.2	5.6[e]
1940	2,906.5	2.6
1941	3,469.1	3.1
1942	3,427.5	7.4
1943	3,598.0	4.5
1944	3,684.1	1.4
1945	3,507.7	3.9
1946	2,811.0	22.2
1947	3,299.5	15.4
1948	3,533.4	30.7
1949	3,019.7	24.4
1950	3,675.7	17.5
1951	3,837.3	34.6
1952	3,307.3	28.3

[a]Source: William H. Lowe, Inland Steel Company.
[b]Increase during year.
[c]Retirements of old equipment.
[d]Expansion of steel capacity and related facilities and acquisition of subsidiaries.
[e]Acquisition of subsidiary.

by one year in 1950. Not only is there question concerning the inconsistency of these turning points due to the possible existence of other lags,[51] but there is also the distortion introduced as a result of the varied motivations which gave rise to them.

51. For example, a decision to invest and a subsequent ordering of capital goods may involve lag intervals which precede the manifestation of the investment but which are the result of output changes or demand changes.

Judgment on the question of investment timing, where based on the expressions of management or as a result of examination of comparatively raw data, can be expected to be rather tenuous, and can at best be the result of what might be termed rational speculation regarding the factors involved. On this basis, however, it may be suggested that decisions to invest are made by management after "assurance" that current new levels of demand will be maintained or that trends will continue.[52] Management thinking is in terms of a one- to two-year interval and in terms of specific types of finished products to meet a specific demand that presents itself. Therefore, some modified accelerator relationship appears to be operative.[53] On the other hand, there may be some rule of thumb practices, such as Inland Steel's, for example, of investing in depression years. There may be specific response to incentives to invest because of tax allowances, and there seems likely to be a continuing emphasis on "rounding-out" of facilities as opposed to launching into fairly elaborate expansion of integrated facilities.

Though the formal aspects of approval may be considered peripheral to our discussion, one example may suggest this dimension of the decision-making process. Eisner cites the following from his interview study:

> The firm is currently undertaking two major capital expenditure projects: basic steel capacity expansion and cost reduction in tin plate and cold rolled sheets. In the first case, in order to feed the finishing operations and maintain our place in the industry a 20 per

52. Departure from "rule of thumb" will occur where alert management recognizes exogenous factors of relevance. For example: "Because of the expected impact of the St. Lawrence Seaway and Chicago area's rapid growth, Inland Steel Co. is projecting an expansion rate faster than that of the steel industry generally. Joseph L. Block, president, says Inland's capacity will rise from its present 5.6 million ingot tons to 7 million in 1962, 8 million in 1967, 11.5 million by 1977." "Get Ready for the New Boom," *Steel, 143* (1958), 99–100.

53. For a discussion of variations in capacity, both secular and cyclical, and with specific indication of the performance of the iron and steel industry, see B. G. Hickman, "Capacity Utilization, the Acceleration Principle," pp. 423–39.

> cent expansion was decided upon. Engineering and operating people were then asked how to accomplish this. After consultation with company executives the proposal was taken to the board of directors and approved. In the second case, the project stemmed from the operating and engineering people. They're always concerned with matters of this kind. They evolved proposals which were taken to the vice president in charge of production. He took them to the president, who gave them back to the executive vice president to be checked and studied. The executive vice president then recommended the project to the president. He will in turn recommend it to the executive committee, supported in this by the executive vice president. It will go finally to the board of directors.[54]

An additional consideration is that concerning the oligopolistic structure of the industry. Comment on this will be left to a later section of this study. However, a point touched on above should be stressed once again: a firm may have to add to estimated area increases in demand, consideration of whether it wants to hold to its past share of a growing area market or whether it may wish to increase this share. The presence of a stronger firm selling in the area and the subsequent reaction by that firm may be an important consideration in the decision-making process.

Now that material has been presented which indicates the more mechanical aspects of accumulating evidence on which expansion decisions are based, it is appropriate to turn to more explicit examples of the range of attitudes held by leaders in the industry. This brings us to a point where factual data are joined to some of management's longer-run goals—or perhaps this should be termed the "enunciated goals" which management expresses. There appears to be considerable confusion among some industry executives as to just what their specific

54. Eisner, *Determinants of Capital Expenditures,* p. 54.

objectives are, or should be. Contemporary business structures, because of the variety and diffuseness of chains of authority, often present an unclear picture of where responsibility for action rests. The crucial issue in this area, for some observers, is reflected in the assertion that "the large corporation does not have an effective machinery whereby business leaders can be readily and continuously held accountable for the attainment of these objectives."[55]

Full treatment of the significance of these questions would carry us well beyond the scope of this study. Nevertheless, our discussion must take note of the degree to which these enunciated goals are in line with the set of so-called objective social criteria referred to in earlier sections. Deeper than this is the question: are the actions taken by firms at variance with their enunciated goals or do they appear to implement them?

Expressions of industry objectives range everywhere from specific industry obligation *to produce* to an elaboration of the broader philosophical involvement with the economy. The former is reflected as a part of a statement on expansion by Ernest T. Weir, Chairman of National Steel Corporation, in 1956:

> Management in the steel industry, to my mind, has the obligation to see *that there are no steel shortages.* In other words, the industry has the obligation to build additional capacity ahead of need. In fact, I believe that at this time the steel industry's total capacity should be 6,000,000 tons higher than it is—as a guarantee against shortages. It is always the obligation of the steel industry to build ahead.[56]

An example of objectives stated in a broader context can be seen in an advertisement of the Jones and Laughlin Steel Corporation signed by Ben Moreell, then President:

55. R. A. Gordon, *Business Leadership in the Large Corporation* (Washington, The Brookings Institution, 1945), p. 340.

56. E. T. Weir, "Steel Expansion: The Problems Involved," an Address before the New York Society of Security Analysts (New York, March 22, 1956), p. 7.

> We're backing a *sure* thing with $210,000,000—the *future* of the United States of America. . . . J & L is a typical American industrial *team*. This *team*, to serve its *customers* and the national security, *needs* the savings of *investors*, the strength, skill and cooperation of labor, the leadership and technical experience of *management*.
>
> We are but one of the many teams that make up our American Industry which is moving with increased momentum toward the goal of *better living* for *all people everywhere*.[57]

And stressing the national security aspect of steel's role in the economy, an advertisement was published by the United States Steel Corporation quoting a speech of Benjamin F. Fairless, then President of the Corporation:

> Americans, of course, don't like to take second place in any league, so they expect their steel industry to be bigger and more productive than the steel industry of any other nation on earth. It is; but what many Americans do *not* know is that their own steel industry is bigger than those of *all* the other nations on earth put together . . .
>
> It is clear that the American steel industry has more than fulfilled what is probably its first responsibility to the nation—the obligation to outproduce any possible combination of aggressors. But what of our domestic needs? How have those been met?
>
> All during the Twentieth Century, the steel industry has maintained an average productive capacity nearly 50 per cent greater than the demands our nation has made upon it. That means that over these years on the average, nearly one-third of all the steel making facilities in America have stood idle.
>
> Yet, in spite of this, it has continued to expand steadily, in every decade—even in the depths of the

57. *The New York Times* (January 5, 1949), p. 15.

> depression when only half our steel capacity was being used and when we couldn't have sold another pound of stuff if we'd taken cigar-store coupons in trade!
>
> . . . No other nation in the world could have matched that record. It is a record that stands as a glorious tribute to the men who make steel and the men who built steel in America.[58]

The bravura of these public pronouncements cannot stand alone as a representation of the managerial point of view on objectives for the industry. A line of inquiry must be pursued which will reveal something more of the motivation behind decisions to act within the industry, and delineate the decision-makers as well as their decisions.

One might suggest that the logical point of departure for generalizations about investment decisions should be the Keynesian marginal efficiency of capital schedule. It is logical, indeed, both in the short and the long run, to expect that a firm will calculate the net annual expected return on a piece of capital equipment, once the replacement costs have been taken care of, in order to decide whether acquisition of the equipment will be profitable. Alternative use of financial resources which would result in a greater rate of return over time would eclipse the planned acquisition, and so the real test of profitability would be that the marginal efficiency of capital would have to be equal to or greater than the going rate of interest. Keynes stressed the advantage of his formulation because it stressed "that the attraction of a capital asset is not its current yield but its *prospective* yield."[59] But as Andrews and Brunner suggest, Keynes

> assumes, without inquiry or proof, that a businessman will meet this problem by constructing a precise estimate of the total expected net yield of capital equipment. . . . The consequence is that Keynes' theories retained the chief characteristic of the older,

58. *The New York Times* (January 2, 1951), p. 67.

59. Cf. P. W. S. Andrews and E. Brunner, *Capital Development in Steel*, p. 10.

> static theories—that the rate of investment will depend upon the rate of interest so far as external influences are concerned. Despite all the splendid passages in the *General Theory* where Keynes discusses the effects upon economic affairs of the uncertainty which surrounds business operations, he yet, when it comes to the point, resolves that uncertainty into the formal equivalent of certainty. It is this which enables the rate of interest to maintain the importance which it has in his theory of investment decisions.[60]

This puts us face to face with the need to appraise the importance of the rate of interest in affecting decisions to invest at the microeconomic level. For if this contention is accepted, one must be carried along into a discussion of monetary questions as they impinge on the general discussion. It has been difficult to find, in the empirical material accumulated—in either recorded statements or in other attitudinal evidence—that the rate of interest does have, in fact, *the* tie to investment decisions that are implicit in the marginal efficiency of capital, interest rate formulation referred to above. It is true, of course, that the problem of "obtaining funds" is stressed often[61] and this shall

60. Ibid., p. 11.

61. Eisner notes that "although there is no direct evidence that interest rate cost, or even cost of capital in general is significant, such considerations must be implicit in the concern about diluting stock and impairing the company's position in the securities market. For a high interest rate might well, *ceteris paribus,* necessitate higher yields to attract capital to new investment in equities. And resultant higher yields on old issues, granted the distaste for 'diluting,' tend to make additional capital look generally unavailable. However, to keep the role of capital supply in proper perspective, we may weigh the remarks of the assistant treasurer, who declared: 'The question of availability of funds usually comes in at the tail end of the decision. After top management has about decided to go ahead, the issue of funds may come up. If funds are short, top management may be asked to take another look, or get new sources of funds.' " Eisner has stated, however, that it is his impression that if a company has a worthwhile project they will manage to obtain the funds to finance it. He states: "One may still argue, it should be noted, that interest rate is relatively unimportant compared with expected profitability and the rest of the complex of factors influencing capital values. See J. M. Keynes, *General Theory of Employment, Interest and Money* . . . especially pp. 147–64" (Eisner, *Determinants of Capital Expenditures,* p. 59). In fact, some investment research indicates that the problem for some firms is one of finding additional outlets for accumulated funds, once that capacity expansion deemed reasonable has been completed.

be developed later on, but the marginal calculations that are suggested in the Keynesian approach are not stressed at the firm level, at least, and do not hold the force that a "classical" Keynesian formulation suggests. The ironic part of this is that it is a part of the Keynesian idea that "prospects" *are* a part of the picture, but as we shall see, the question of expectations is complex, and embraces a range of variables which (though it may include the rate of interest) reaches further, beyond this one consideration.[62] At the conclusion of their study of British steel, Andrews and Brunner state:

> No evidence has been found that the decision to make or not to make a given piece of capital expenditure has been affected by the level of the rate of interest or by variations in that level. . . . The point is that changes in the rate of interest seem, not only in United Steel but in manufacturing industry generally, to have had no effect upon currently planned capital expenditure, and that this is likely to hold good for any changes in the rate of interest which are at all likely in practice. If this is so, then the economic theory of investment decision has to be recast in order to remove the traditional emphasis on the rate of interest as a key influence in those decisions.[63]

62. Andrews and Brunner quote two letters they received from businessmen which touch on this: " 'Really you know, the factors which decide any policy of capital expenditure are human: greed, love of power, jealousy, anger, desire to benefit the country or industry or company employees, all inextricably mixed, but one and then another becoming paramount at different times and influencing decisions.' " and " 'I often feel, however, that in economic arguments the effect of personality is overlooked. Some businesses grow at the expense of others because they are managed with more skill and forcefulness. No businesses grow of their own accord and, equally, amalgamations do not take place unless some personality with vision sees the advantage that lies in that course of action.' " (Andrews and Brunner, *Capital Development in Steel,* pp. 14–15).

63. Ibid., pp. 347–48. Note that, as was stated above, there is nothing in Andrews and Brunner to indicate that management ignored the terms on which finance was to be obtained. On the contrary, there is evidence that there is a strong presumption that steel management considers very carefully both that the "cheapest" method of financing is sought and also that the decision to resort to temporary financing as opposed to "long-dated finance" are matters of great concern.

If one thus rejects the classical Keynesian formulation as the most appropriate ideal characterization of the decision-making apparatus, because of its over-emphasis on monetary considerations and its essentially static character, what alternatives may one turn to? A survey of the literature on expectations reveals that that which is referred to as a "naive model" is often a point of departure; it is essentially the assumption that present trends will persist, that entrepreneurs will base calculations regarding the future on a continuation of present trends. This is at once a compromise with the logic of the uncertainties businessmen face and a tacit assertion of the persistence of "stable" economic conditions. It suggests that there is a pattern of habitual behavior in business decision-making, and that entrepreneurs may well reject planning for a rapidly expanding market in favor of using linear trend projection.

A modification of the continuation-of-present-trend idea is the approach in which firms follow a pattern of imitating action which in the past has brought success. Alchian discusses this, and lists the following as reasons for adopting such a line of action:

> (1) the absence of an identifiable criterion for decision-making, (2) the variability of the environment, (3) the multiplicity of factors that call for attention and choice, (4) the uncertainty attaching to all these factors and outcomes, (5) the awareness of superiority relative to one's competitors is crucial, and (6) the nonavailability of trial-and-error process converging to an optimum position.[64]

He goes on to suggest that "imitation affords relief from the necessity of really making decisions." He points to the element of natural selection which allows those who depart from imitation at the "right" time, as well as those who stick to past patterns when this is the wise choice, to achieve successful results. The survivors are thus vindicated in their line of action. Though an effective ex post representation of an evolving structure,

64. A. A. Alchian, "Uncertainty, Evolution, and Economic Theory," *The Journal of Political Economy, 58* (1950), 216.

Alchian's attempt at analogy with a biological phenomenon still leaves us without a basis for anticipating why the decisions to stick to old forms or to embark on new ones are made.

There are two grounds upon which it seems logically valid for an observer to use the continuation-of-present-trends assumption. The first of these, of course, is in the case where this condition closely approximates the realities of the case; the second seems to be where the degree of aggregation involves coverage of a fairly diverse group of firms or industries, and where this assumption serves as a reasonable representation of action on which certain functional relationships are based. We must recognize that in the second case, extensive qualifying information must accompany conclusions drawn from the model. Otherwise policy recommendations which have their impact on individual firms may be completely inappropriate, as the *particular* firm or industry may be precisely the one which has been a deviant element within the aggregate, and such policy formulations would be completely inapplicable. Given the purposes of this study, generalizations concerning such functional relationships, though logical and desirable in aggregate model building, will not satisfy the objectives we seek and we must turn our attention elsewhere.

Two initial premises will serve: (1) that the context is dynamic, i.e., that there is change and uncertainty exists, and (2) that perception intervenes between reality and decisions to act. Let us look at the second of these premises and turn our attention to the question of cognition.[65]

Some writers have taken the position that profit maximization is an adequate representation of motivation for purposes of generalizing economic action. This maximization principle is not empirically validated in individual cases, however, and formulations which suggest a tending-to-equilibrium pattern are inappropriate in a dynamic context. Simon suggests where these conclusions lead:

65. This, of course, cannot be an exercise in epistemology. We are not concerned with "the nature of reality," but rather with reality perceived and translated into a basis for action.

> As we move from maximizing theories, through simple stochastic learning theories to theories involving pattern recognition our model of the expectation-forming processes and the organism that performs it increases in complexity. If we follow this route, we reach a point where a theory of behavior requires a rather elaborate and detailed picture of the rational actor's cognitive processes.[66]

He goes on to point out the need for a model that stresses that

> alternatives are not given but must be sought; and a description that takes into account the arduous task of determining what consequences will follow on each alternative. . . . The decision maker's model of the world encompasses only a minute fraction of all the relevant characteristics of the real environment and his inferences extract only a minute fraction of all the information that is present even in his model.[67]

We must be careful to realize that our interest in this decision-maker is not in terms of his social role but rather in terms of the ramifications of his actions in economic life. Having said this, however, we still must look at the decision-maker in a social context. This is the central point: his perceptual field is conditioned by his *role* as an economic actor, and we must try to get a picture of him in these terms. Inquiry must move in the direction of characterizing the cognitive link in order to understand the decision-making process. Perception may be thought of as conditioned by role fulfillment, as a kind of deflator in relating objective reality to the thought processes which lead to decisions to act.

There is another factor, a complement to role perception: the perception that is held or is experienced with regard to the *environment* per se and the subsequent interaction of these two factors. As Simon has indicated

66. H. A. Simon, "Theories of Decision-Capital Making in Economics and Behavioral Sciences," *American Economic Review, 49* (1959), 272.
67. Idem.

> a role is a social prescription of some, but not all, of the premises that enter into an individual's choices of behavior. Any particular concrete behavior is the resultant of a large number of premises, only some of which are prescribed by the role. In addition to role premises there will be premises about the state of the environment based directly on perception, premises representing beliefs and knowledge, and idiosyncratic premises that characterize the personality.[68]

This expresses—in somewhat special terminology—the mandate to move beyond the profit-maximizing objective for these men and to include such perceived goals[69] as the desire for security, the continuity, growth, prestige, and power of the firm and the degree of identification of the individual decision-maker with the firm, as well as the desire to make contributions to social welfare and the desire for social approbation, and even, as a limit, the "participation in an absorbing game." Discussion of the perceived role suggests that the basis on which the individual "identifies" may not only be different from that typically assumed under the profit-maximizing frame of reference but that in fact there may be "multiple identification." For example, the decision-maker may be more anxious to achieve success within the organization than success for the organization in the market place, two goals which may be at variance with each other.

This leads us to the assertion that the conditioning, training outlook, and organizational structure within which individuals operate all will have bearing on the kind of decisions forth-

68. Ibid., p. 274. Simon goes on to point out that "with our present definition of role, we can also speak meaningfully of the role of an entire business firm—of decision premises that underly its basic policies. In a particular industry we find some firms that specialize in adapting the product to individual customer's specifications; others that specialize in product innovation. The common interest of economics and psychology includes not only the study of individual roles, but also the explanation of organizational roles of these sorts" (ibid., p. 278).

69. Cf. J. P. Miller, "Contributions of Industrial and Human Relations Research to Economists' Theory of the Firm," Proceedings of the Eleventh Annual Meeting of the Industrial Relations Research Association.

coming, and also on the changing pattern of these decisions as factors change from person to person, from firm to firm, and from one point in time to another. This may appear to be rather naïve sociological questioning, or perhaps suggests an exercise in social psychology. On the contrary, our interest in such information rests on a desire to answer the specific question tied to policy formation in the macro-economic area treated in earlier sections of this study: how should one evaluate the possible discrepancy between the objective needs of the society and what may be expected to be forthcoming from individual entrepreneurs, acting according to a certain motivation pattern fixed from within the industry?

THE INDUSTRY VIEW ON CAPACITY NEEDS VIS-À-VIS THE AMERICAN ECONOMY

Among the various approaches that can be used in trying to answer the foregoing question, the intimate case study method is one of the most promising. However, the word *intimate* is significant, and suggests a kind of psychoanalytic intensity and persistent observation which transcends the scope of this study; more than this, it would involve an outlay of time which would not be compensated by specifically relevant insights greater than can be approximated through more economical means. There is also the question of whether this type of study would be appropriately undertaken by an economist as part of an economic study. An alternative that suggests itself is a representative sampling, handled at a perhaps more superficial level, based on a kind of cue system. If this approach is to be justified, and to succeed, it must be supplemented by some kind of correlation to observed actions—that is, by a test of whether decisions actually made are consistent with a set of generalizations derived from fairly limited observation of the so-called "representative" entrepreneur. To accomplish this, we will use not only a summary of recorded statements by industry executives over a period of time, but also a set of observations of a particular episode in the history of the industry's pronouncements in this area: we will draw on discussions of the expansion needs of the economy for steel, and the industry position regarding them in the period 1947–50.

As the horizon of interest in this study is one which reaches relatively far into the future, specific quantitative estimates used by decision-makers do not provide a useful point of departure for the evaluation of attitudes and action. In the steel executive's world, an anomalous situation exists that forces him to move adrift of the moorings which short-run tight analytical models might provide him; he must plan in the face of longer gestation lags. And so even though he may have confidence in and sympathy toward the use of precise and carefully formulated systems of analysis, there is nevertheless a knowledge lag which forces him to lean more and more on past experience, rules of thumb, and the elusive ingredient which is usually referred to as expertise.[70]

What follows is a report on that phase of this study which deals with the so-called behavioral side of the picture. The material consists of two lines of inquiry. One is a report on interviews, conducted with a sampling (not a sample) of high-level decision-makers in the industry. These opinions have not been taken at face value, and will be presented in the form of a running account with interpolated comments (subjectively conditioned as they undoubtedly are) to link the expressed ideas and opinions to the more "theoretical" variables referred to in earlier sections.[71]

It might be mentioned parenthetically that the line of questioning did not include what is often referred to as organization or communication theory, i.e., the apparatus through which planning is channeled up and down. Emphasis was more on the businessman's profit-making role, responsible to whomever he feels responsible, and his so-called social role, in whatever form

70. This statement of the state of the "forecasting arts" today is in no way meant to detract from the hoped-for improved power of demand analysis, suggested in Chapter 4.

71. The second line of inquiry (which follows, below) is an analysis of published pronouncements of leaders in the industry over an extended period of time.

he conceives this to be. Many peripheral (that is, peripheral to the main issue of concern) opinions were developed in the course of these interviews, some of which have been included in the report of the discussions.

Several remarks must be made regarding the direct interviews and the basis upon which they were undertaken. An attempt was made to use an open-ended interview, for several reasons. First, in an interview without a pre-determined structure, it was hoped that the executive's own ranking system of the importance of issues would come to the fore. He would be permitted to set up the hierarchy of topics so that, from the interview, the implicit weighting of issues of interest to us in this study might emerge. Second, we might also find a basis for drawing conclusions on a number of qualitative factors, e.g., personality type, intellectuality of approach, degree of vigor, conservatism, public relations consciousness, sensitivity as regards the industry's role, of suspiciousness of the academically-originated inquiry, of self-confidence consistent with the particular rank of the individual in the organization, and so forth. As these interviews were approached without any preconception of their results, it was hoped that the lack of structure would aid in keeping this aspect "uncolored." There is often a great danger, in the direct interview, of the interviewer leading and channeling the course of the discussion so that his preconceptions manifest themselves in the report. My objective here was not to come away with a tally sheet of responses to a specific set of questions, but to obtain an impression.

Given the set purpose of the interviews, therefore, there seems to be no need to apologize for the impressionistic quality of these records, which reveal, to some extent, the individual's personality as well as his expressed views on the issues discussed. The reader, however, should be alerted to the phrase "expressed views"; there is danger in assuming that more has been revealed than is actually the case. In what follows there will be no attempt to disentangle the "real" views and the unconscious motivation of the individual, from what he expresses; and there need not be. The fact that a view *is* expressed indicates conscious motivation on the part of the subject to convey a pattern of behavior, to present the personality face that is presented. This, in and of itself, is indicative of a significant facet of the total personality. Of course, one must allow for subterfuge and for public relations postures, as we attempted to do in the interpretive parts of the interview report. It is true there is a danger that the observer may be carried away with his own ability to divine the "real man"; however, for the use to which the information is to be put here—as opposed to those uses which might command the attention of a social psychologist—this approach seems justifiable.

The general format for these conversations was as follows: the person interviewed was told that a study of the steel industry was under way, that the study would be "atypical," in the sense that the author was not making a comprehensive survey of the iron and steel industry in which the resources, technology, levels, and types of output of the industry in question would be detailed, nor would he preoccupy himself with problems of monopoly and competition and the impact of market structure on prices and output in the industry. Nor was this study particularly concerned with questions of integration, either for reasons justified by economies of scale, or the need to assume positions of power in particular sectors of the economy. The person interviewed was told that these issues, which he might expect would be central to the typical steel industry study, were peripheral as far as this particular inquiry was concerned and, in contrast, this study stemmed from a particular policy question whose origins were in an actual historical occurrence, and that its objective was to gain some insight into the way a particular policy problem (viz. the 1949 controversy) should be resolved, should it recur as a significant problem in the future. The intention of this introductory statement was that the interview would lead into questions relating to investment decisions, access to finance, and the recognized impact on decision-making of interdependence, and from this into questions of the leadership in the industry, the recruitment of new personnel for executive positions, the reliance on guidance and advice from individuals outside of the industry, and the executive's characterization of what makes an effective executive. It was hoped that this coverage would elicit statements providing a basis for discussion of the personality types which hold important decision-making positions within the industry, as well as about their views on substantive matters. Finally, the question of the role of government was developed, at some natural point in the discussion and, using the issue of 1949 (and the strong position of the industry opposing further expansion of capacity) as a point of departure, an attempt was made to determine views on this subject and an assessment of the situation since the Korean War. It also was hoped that some predictions would be forthcoming with regard to (a) future expansion of capacity and levels of output in the industry, (b) the type of personnel and leadership that might be found within the industry in the future, (c) future implications of government interference in industry activity, in whatever terms the individual desired to put it, and (d) the evolving pattern of industry structure vis-à-vis monopoly and competition.

Of the interviews conducted, four will be reported in some detail. They were conducted at the offices of the executives in question and involved chief executives or directors of the corporations involved. In some cases, subordinates or associates were present; their contributions are reported as a part of the respective interviews. Material drawn from conversations of other individuals in the industry, but not treated in these specific interview reports, will be covered implicitly in the summary remarks which appear at the end of this chapter.

INTERVIEW A. After a brief descriptive discussion of the exchange before Congress in 1949–50 concerning the adequacy of steel capacity, this executive suggested that the differences apparent in estimates of future demand made by the various individuals who had studied the question were not unique to the hearings, and that as a matter of fact one would find differences within the industry, even within a firm, in estimating future demand and hence what would be considered adequate capacity levels. The point was made that in the case of differences within a firm, these would of course be resolved before any action was taken to alter the capacity of that firm. But in the industry, he asserted that it was precisely the differences in estimation regarding the future that allowed for competition. He stated that without differences in expectations, there would be no competition. This raises, in the economist's mind, of course, interesting questions as to just wherein the well-spring of competition lies. In any case, this executive pointed out that his firm's estimates had been very close to the demand which in fact did appear—that is, within reasonable tolerance limits.

In reply to the question, whether better tools of analysis might not be developed, he said that the techniques of analysis his firm employed were undoubtedly equal to the best known to date anywhere in the industry. There was the suggestion that automobile producers perhaps had better data, because of automobile registration records, but his firm did as much as was possible with the data at hand. It seemed that he did not really understand what was meant when reference was made to "models of the economy" and to the fact that there might be a deficiency in the type of analysis used by the industry, and so the point was dropped. From his remarks, it appeared that as far as forecasting was concerned, his thinking was clearly based on some kind of market analysis approach. He suggested that the kind of planning horizon which we were discussing was the one most difficult to forecast; the very short run (and he mentioned a figure of three months) or the very long run presented an easier task, as far as projecting figures was concerned. He emphasized that the intermediate period presented the truly difficult problems.

In connection with questions on what had brought forth the increase in steel capacity after the Korean War, he made the point that as far as his company's undertakings went, they had been "in the works" (were on the drafting boards and had been under way) earlier, before anything was known about the Korean needs—and if the Korean situation had some effect, it was to bring abcut a larger expansion of capacity, i.e., shooting for a larger target, than would have been undertaken in the absence of the emergency. He said that when one looks at the time span associated with the Korean situation, the increase in the amount of steel produced was so rapid that much of it must have been planned for before the Korean War.

Then, turning his attention to the pre-Korean controversy, he suggested that possibly what underlay the testimony of some of the individuals who accused the industry of falling short of its responsibilities (though they were not that explicit in the testimony) was a predisposition on their part to look forward to more and more government control of basic industry. This, it was suggested, is what may have led to their presentation of figures which in turn resulted in extremely generous predictions of demand. Our brief exchange on this subject left the impression that this executive felt that part of the drive behind those who opposed the industry's position was related to their interest in more central control of the industry by government. The implication was that here was one more attempt to badger the industry, to put the industry on the defensive, and to provide ammunition, in effect, for a campaign to take over the industry.

We then turned to another point which this executive felt was a part of the over-all picture. He made the point that there was *real* competition in the industry. Contrary to what public opinion might suggest, competition in the industry was responsible for much of the stimulation to expand. His firm, he assured me, certainly was constantly alerted to maintaining or continuing to take hold of the expanding market, i.e., to satisfy the expanding demand. He suggested that as competition explained in part some of the differences in the estimates that appear from time to time from various sources, so, too, different firms follow different paths in getting new business and react differently as a result of their differing estimates of future conditions.

When asked whether optimistic estimates on the part of a competitor spurs his company on (that is, do they get on the bandwagon and want to expand also?)—or do they, on seeing others expanding, read into these programs the likelihood that the industry as a whole may be overexpanded (that is, do they pull in their horns and cut back on their planning?)—his answer was that either might result, depending on the circumstances. In the period around the Korean War, he attested, his firm felt that some of the individual estimates of companies and what they planned to do added up to an aggregate expansion (he suggested a figure somewhere near fifteen million tons) which seemed unreasonably liberal. But then he went on to acknowledge that, in fact, it proved to be quite right. When asked whether this was something of a self-fulfilling prophecy, he turned the answer around to say that one is bound to find that estimates, though falling short, may well be expected—even though off the mark—to be, nevertheless, within so-called tolerance limits found in any forecasting operation that has elements of uncertainty. It was, he emphasized, this situation of uncertainty which gave rise to the differences in various estimates. Note that he chose to point to uncertainty rather than faulty forecasting techniques.

He mentioned the testimony of one steel executive at the time and pointed out that with that individual you do see an example of someone of quite a conservative bias. His own firm had never felt that things would be so dark and demand so easily filled as was suggested in that testimony. In this part of the discussion, one finds an awareness among executives of differences in their relative predispositions toward "conservatism." Perhaps it would be better to say their differing attitudes toward the future: i.e., that distinctions do exist and that the executives are conscious of them. He noted in passing that it was likely that there would be gaps in the testimony of certain companies, because some companies had a policy of not making their estimates public or expressing them quantitatively for Congress or anyone else.

When questioned about whether he thought energy should be expended in an attempt to bring these differing estimates of future demand more in line with each other, he gave the impression that he believed pursuing this would be worthwhile, but that it might actually be of little significance. He seemed to react unfavorably to my pressing the point of the discrepancy in estimates as a fault of poor marketing analysis techniques; he felt that these tools were in good shape. Where differences arise either between government economists' estimates and those of the industry, or where differences within the industry arise as a result of the various firms' perspectives or techniques of analysis—that is, where there are "honest differences because of methods of estimation or an inclination toward conservatism or something along these lines,"—these differences, he contended, fall pretty much within the realm of tolerance, pretty much within the realm of what discrepancies must be expected when one looks into the future.

He did suggest, however, that some differences are contrived or deviceful. Where there is motivation (and he indicated "certain circles" outside of the industry) to promote disharmony and apparent conflict of interest, then the problem is to deal with such pronouncements in terms of their source and the motivation behind them. In this case, he contended, statistics are fitted to the preconceived goals of the analyst and are rationalizations of a position rather than resultant upon techniques of estimation. Where this is the case, he felt it was not arguing to the point to become preoccupied with estimation problems, because larger issues of policy lie at the bottom of the question.

Putting his feelings in a "positive" way, he asserted that in a truly competitive situation, individual firms will be forced to do the best they can to estimate the future and that these same firms, then acting upon their best intelligence, will inevitably act to expand capacity so that they may retain their share of the market. He pointed out that when one views the industry, one sees a change in the relative market shares of the various firms in the industry over time and that this should be construed as evidence of the dynamic competitive situation in the industry. He asserted that when one talks about the industry as a whole failing to meet needs, discussion has moved into a realm of social and political policy (implying issues of capitalism and socialism, though he did not identify these two forms of organization as such) dominate the scene. Here is an example of how readily discussion can move from the specific industry situation to the industry's role in society. The executive will frequently take this route to include questions of social goals in his interpretive remarks.

When asked to comment on the possible economies to be realized in embarking on new projects of large proportions (integrated works at a new site), the executive indicated that once an expansion project got under way, the firm often found it was able to realize economies or output greater than it had originally anticipated and that this might be considered part of the idea of a self-fulfilling prophecy. He indicated that increases in the productivity of machines can, of course, be embodied in a new plant, but also that facilities, when pressed to operate at levels beyond their calculated levels, do indeed at times come through. This suggests that there apparently is leeway, and that efficiency yielding greater output than may originally have been anticipated or planned for is a possible by-product of expansion. Once an expansion program gets under way, economies appear in unexpected places, and the over-all result is an increase in effective capacity greater than what had been anticipated. In this executive's view, this point served to underline the problem of tolerances on estimates, not only on the forecasting of demand side, but on the production side as well. It serves also to underline the problems associated with quantitative planning for capacity expansion developed in Chapter 4.

He touched on the special problems in creating balanced growth facilities in industries such as steel because of heavy capital outlays. He also touched on the indivisibilities question. He cited financing as a very important factor in the expansion question, and said that there is no doubt that accelerated amortization had enabled the industry to expand at a faster rate and achieve aggregate capacity levels in excess of what would have been possible were this tax benefit not available. He also emphasized that the industry did respond in a patriotic way in times of emergency, and that an indication that his firm was not completely profit-oriented could be seen in their provision of some nominal cost facilities for the government at such times. This, he stressed, is evidence of the conscience and sense of social responsibility that firms do have.

To report some partial impressions of the personality of this executive: he was a very incisive individual, extremely self-possessed. He took the initiative in the conversation. He apparently saw this as an opportunity to persuade an academician on the goodness of the industry and the rightness of its way of looking at issues. But at the same time, he was perceptive enough not to hurl clichés; rather, he was willing to let specific questions be raised and to answer them quite directly. He apparently felt it desirable to be as forthright as possible and not to pull any punches as far as the questions went; this was in part a reflection of the temperament of the individual and in part a sense that this session provided an opportunity to accomplish some public relations objective. A most striking impression was that this individual, in charge of policy formation for his firm, felt the need to take valuable time with others, outside of the industry, in the hope that a patient, rather straightforward discussion would tend to influence the image imparted.

INTERVIEW B. After a brief outline of the reasons for undertaking this study, the executive was asked how he might explain the difference in the behavior of the industry both before and after the outbreak of the Korean War. He indicated that he had been involved in industry efforts to plan for meeting the demand from the military establishment as the Korean War broke out. The industry, he indicated, saw that there was a possibility of running short on supply. He had spent time in Washington in an effort to marshal the industry, to set up quotas and estimates of expansion needs on behalf of the industry. At one point, fifty men were working on such estimates. He said that he found that in his experience in Washington, both then and during World War II, the most serious problem faced by those in this work was to acquire numbers; that is, they had difficulty persuading officials in high places to be specific as to the kinds of products they needed. He emphasized that it was not sufficient to talk in terms of so many ingots of additional output, but at least from the industry's point of view it was necessary to know whether the specific need was for sheet, bar, strip, plate, or whatever. He indicated that he felt that the response of the industry at the time was simply and merely reaction to contingency needs, to the crises that arose from Korea. He acknowledged that although accelerated amortization had some effect in accelerating the pace of some of the expansion, the main accelerating force had been the desire of individuals in the industry to meet the demands of the economy. He elaborated in some detail the difficulties in planning and the complications of engineering which require plans to be made well in advance and which involve considerable uncertainty because of the special design problems associated with producing special equipment. He cited as an example the elaborate number of motors that must be designed and produced for any new rolling mill installation.

In referring to a significant expansion project of his firm, he pointed out that the project had not been the end product of some formal planning procedure.[72] There are various opinions as to whether a less formal procedure is more desirable; this executive felt that it was, and that a continuous line of communication among the decision-makers is essential. He showed the need for constant checking to see that one phase is consistent with other phases of the plan and the needs of the various parts of the organization. In other words for him a completely fluid mechanism would be most fruitful.[73]

72. It is clear, however, that the actual mechanics of the planning procedure vary from firm to firm, and this executive's firm did seem somewhat atypical (cf. earlier material on planning from Eisner). Whether a formalized procedure is involved or not, it was clear from this interview and others that the initial impetus to expansion may well come from a variety of different sources. For example, on some occasions individuals in the sales force may initiate talk of expansion and on other occasions it may be a man in charge of operations. At times individuals involved with the international activities of a company may trigger the planning sequence. In any event, when plans are drawn for expansion and when engineers have submitted drawings and items are costed, they must inevitably be cleared through a sales analysis division.

73. Other opinions expressed suggest that this leads to bureaucratic inefficiency and could only be viable under certain special conditions, particularly when directors and executives are in very frequent communication.

As in the preceding interview, discussion then turned to problems of the development of balanced facilities. Emphasis was placed on "the fundamental error" in aggregate output figures for setting targets for industry capacity. It was stressed that unless there is a breakdown in terms of product mix, the likelihood of obtaining meaningful demand and capacity targets would be limited. In this connection, it was pointed out that when decisions are made in a formal way, it is necessary to deal with a time horizon explicitly stated, and this is difficult as target dates must be varied intelligently, depending on what facilities and what phase of operations are being expanded. He cited, for example, the development of ore resources wherein a firm might have to plan the development of fields some twenty-five years in advance. It conceivably could be developing resources that never would be put to use. On the other hand, it is impossible to ignore such a factor until the day of reckoning comes upon the firm, because then, of course, it would be caught short. A striking feature in this discussion was the continual emphasis on the need for a firm to be preoccupied with avoiding bottlenecks—the point that the industry, of necessity, is concerned with having adequacy as far as its resources and its inputs are concerned. The industry is very much aware, he asserted, of the problem of bottlenecks. The best that could be elicited with regard to planning horizons was the assertion that the planning horizon must be related to the particular facility under consideration—that it is a kind of ad hoc situation.

The discussion next touched, implicitly, on the problem of economies of scale. The subject noted that if a piece of equipment were to be made which would enable the firm to produce ten thousand ingot tons of additional output and the firm had demand for only five thousand tons, the question of whether or not to go ahead would have to be faced. The firm would probably go ahead, he pointed out, but with the realization that this facility would have to be operated at less than full capacity. When it was suggested that this could not be done for long, particularly if competitors were in a position themselves to operate near full capacity, he acknowledged that this was true, but that if the decision were made to take the jump and move ahead with this indivisible capital outlay, it would be based on the assumption that the firm were facing a dynamic economy and that demand would catch up. He added that, indeed, taking such action would be a mark of *dynamic management.* It was dynamic management which would go ahead with such an acquisition and which thought in these terms.

In regard to expansion planning, he returned to the bottleneck problem and repeated his emphasis of the balanced growth problem. This was expressed in terms of the difficulties a dynamic firm found in keeping its facilities in balance because it was always moving ahead with first one phase of its facilities, then another, and so forth. This was inevitable, he indicated, because of the fixed and usually large jumps in capacity associated with a new piece of equipment. He referred to the experience during World War II and Korea and stressed the government's lack of understanding of the multi-product character of the industry. One significant contribution to reconciliation of the discrepancy between the government view and the industry view, he suggested, would be a detailed analysis by government economists of future demand *by product.* He referred again to his inability to get, when in Washington, a pinned-down estimate from the military of need for steel by type of product. He acknowledged that the estimate finally came during the Korean crisis, but it was difficult to obtain.

In talking of the Korean period, he said that the industry responded to the political crisis without giving thought to the possibility of falling demand after the war. He said, in fact, that his firm could not have waited for the government's final estimates and that the firm actually went ahead to expand. He acknowledged that accelerated amortization did play a role in the decision.

The discussion then turned to the question of personnel. It was pointed out that economists typically talk in terms of risk avoidance, and that the industry is characterized by some as manned by risk avoiders. What was his reaction to this? He said that this was not the case, that management was very anxious to move ahead and take advantage of expanding demand conditions. When it was suggested that there might be some divided loyalties and that at one time steel leaders had declared "The public be damned!" he pointed out that then they were the proprietors. He stated that Carnegie was interested in himself, and "had Charlie Schwab out there making money for him," whereas today there were thousands and thousands of stockholders, and today management was very conscious of its obligations to the stockholders.

At the same time, management is very sensitive to its obligations to the society as a whole. He cited as evidence of this fact that in his firm, for example, one of the vice presidents—an individual who sits on the board of directors and has status equivalent to the vice president in charge of operations—is the vice president in charge of public relations. Putting it this way suggests, to this writer at least, that the firm under discussion sees the role of the public relations officer as one of "selling" the company's policy, whatever it may be, to the public as in the public interest. There are undoubtedly various reasons why firms elevate the public relations man to this high status. Some do so for the purpose of convincing the public of the position's importance and of convincing the man himself of his importance in the organization so that he does have some weight in his public pronouncements; others have the view that he must have status so that he can have impact on other personnel in the firm and so that he can act as a kind of public conscience for the firm. As it is true that this varies from company to company, it is probably also true that within some firms the public relations officer's role has two parts: that of the conscience of the company as well as that of its Madison Avenue representative.

Regarding differences in the type of individual in management today as contrasted with earlier years, this executive suggested that the early administrators were quite different. He cited as an example a man in charge of a works some time ago, who expressed the opinion that "chemists would be the ruin of the industry." His point was, of course, that all this had changed and that today executives are acutely aware of the need for a scientific approach to production. He noted that the men today have college backgrounds. Each year a special force from his firm goes out to recruit young men from colleges, and many (in his firm's case, 300 or so) are brought in for interviews. The personnel department is given an estimate of what specialties are needed and in what numbers (perhaps 20 engineers, one economist, etc.). All of the chosen men are put through a basic training program, lasting a year (sometimes longer) and then move along and up, some ultimately reaching the top levels of management. He indicated that there were individuals in the present top management who had started in such a program some twenty-five years ago.

He stressed that the main thing, as far as personnel was concerned, was that they were all steel men, that they knew the work, that they had been in the industry from the bottom up, that it was a continuing policy of the firm to develop their own people through the ranks, and that the experience and conditioning these men received over the years made then uniquely suited to take the positions at the top and ultimately to make important decisions. That personnel might be moved from government to industry or that the government ever would be in a position to go into the steel business was fallacious, because their men did not have the kind of cumulative experience the industry people had acquired. It could not be obtained in any other manner than the way it is done in the industry; that is, by men being in and belonging to the industry. This point was made with considerable force.

In turning to the relationship of the industry to government, the executive seemed to restrain himself. He seemed to hold relatively bitter feelings about the desire or ability of government to comprehend the position of the industry. For example, when asked if he would advise me to extend these interviews to include individuals at the Department of Commerce, and whether it might be expected that public officials would be antagonistic to the industry, he shied away from the question. He suggested that he could guess what would be found there and that it probably would be best to "find out for yourself." The impression was given that I would find the so-called specialists within government not particularly perceptive as far as the actual problems faced by steel executives were concerned.

An over-all impression of the interview was that in this individual, a man whose insights obviously contributed to policy formation of his firm, there was somewhat more tolerance and understanding of the professional economist than in many others in the industry, an individual who was quite conscious of public relations and recognized that he was talking to a person viewing the industry from the outside. This brought forth a somewhat guarded representation of the industry position. He seemed a little surprised by the direction taken by the discussion, disregarding, as it did, more typical emphasis on output and technology questions. The industry, so often confronted on the price-profit side, seems to be well prepared to answer questions on this topic, and has not, at least recently, had to explain its position regarding problems in the context in which this discussion placed them. This appeared to have some effect in setting the tone of the interview. On the positive side, the conversation contained fewer already-developed pat answers to questions than would have undoubtedly been the case if the material had been on price-profit questions.

INTERVIEW C. This interview concentrated on the planning horizon of the executive's firm and its decision to undertake a major development project. He first indicated that his firm had a long-range plan of twenty years that was constantly being revised and a more precise plan for five years. He elaborated on the development of a major project and made the point that additional capacity was under consideration as early as the '30s. The onset of World War II delayed further consideration of a major expansion. During the war, the firm had participated in the government's efforts to build or round out facilities, and it was not until after the war that it once again surveyed the scene in terms of regional markets and penetration of new markets. It was a combination of this re-survey and the more recent developments of new ore discoveries outside of the United States that shaped the firm's plans for a post-war facility.

His main point was that the decision to go ahead with a new, separate, and integrated works on a new site was not primarily to add to his firm's aggregate output but rather its desire to gain ground in a competitive arena where over the years it had been losing ground—to recapture some of that market by serving it with a new facility. Competitors had established facilities in a new location. This fact, combined with increased transportation costs from older sites, forced his firm to expand or face the alternative of being gradually pushed out of the picture. Hence, post-war discussion of this facility was well under way before the Korean crisis, and reference to "changed industry attitudes" after Korea about expanding capacity was an inaccurate representation of the sequence of events.

The conversation then turned to the locational factors that planners in the firm must consider. Technological trends make mandatory the selection of sites with an abundant water supply and which make it possible to ship scrap from the market and products to the market easily as well as to receive ores from ocean shipping. The executive commented that basic locational logic contradicts the efforts made in New England to establish facilities where the market was just not sufficient, particularly in view of the fact that New England demand was for special steel. He agreed with the view that the desire for a mill in New England might be analogous to the drives of certain underdeveloped countries for steel capacity, even though strictly economic conditions would advise against it. But his point was that intelligent management would constantly build into its plans shifts in the locational determinants of the industry.

This firm relies heavily on the fact-finding efforts of its experts. In discussing the decision-making apparatus, the executive indicated that men from sales, production, and so forth, present their plans to the top executives. The board of directors is the agency that decides, largely on the basis of the fiscal feasibility of the project proposed. Members of the board tend to take on faith the projection of demand, needs for certain types of facilities, and so forth, and see themselves as the arbiters of "what is possible," taken as given that the firm "always wants what is best for the economy." It was indicated that accelerated amortization did have effects, but that these could be seen only in very recent programs, and did not lie behind his firm's initial motivation to embark on the project discussed above. In commenting on problems of the rate of depreciation, a good many words were devoted to the fact that taxes were not consistent with replacement problems, because original cost calculations were inappropriate for replacement need—a matter frequently brought up in discussion with others in the industry.

The general impression given by this executive was that he was an old hand at umpiring and liaison work, an individual able to evaluate possible lines of action that would lead his firm on a successful course. He seemed to be very much oriented to finance, with a less clearly defined image of technology (although impressed with late developments) and a knowledge of markets in the aggregate rather than in terms of a face-to-face customer relation. From this discussion, in contrast to some remarks in Interview B, we can see that it does not seem mandatory to think of the ideal steel executive as an individual who has "dirtied his hands" with making the product or as necessarily having been involved in the sales or engineering end of operations—that administrative experience or legal experience brought from outside of the industry might well satisfy the needs of effective leadership, if accompanied by a sense of identification with the firm and industry, high intellect, and other effective leadership qualities. This has implications for a theory of leadership: it suggests that if the decision apparatus of a firm is a mechanism, as it seems to be in many in the industry, wherein top management sits in judgment and wherein staffs, composed of what might be called "steel experts," feed material to them, then the type of expertise needed is of a more general and perhaps less technically-oriented variety. Discussion here and with others suggests that a more general set of personality and intellectual attributes defines the limits of expertise, rather than a life of "working in the mills."

INTERVIEW D. After a review of the nature of this study and reference to the 1947–48 hearings on capacity in the industry, the executive acknowledged that his firm had been involved in the preparation of some figures that were indeed conservative figures, and moreover that this was consciously done. The company was very much aware, he noted, of the fluctuations that existed in the market for their products, and their biases surely led them to think in terms of conservative estimates of growth and demand rather than something else. This, of course, confirms the view that the thinking of individuals in preparing that pre-Korean testimony was influenced by their pre-dispositions on the subject, and in the case of this particular discussion at least, there was no reluctance to indicate that such considerations were then operating. The use of per capita consumption of steel as a fixed factor, and the mere projection of it, were acknowledged not to be very sophisticated methods; it was recognized in retrospect that the per capita consumption of steel had increased tremendously and that this was in part, at least, the explanation for the increased consumption pattern that exists today.

In inquiring about what does influence thinking and planning for the future, it was clear that the people in this firm relied heavily on reports from salesmen. They seemed to pride themselves on the good working relationship between their sales force and their customers, and felt that one of the by-products of this good relationship was a fairly sensitive and accurate representation of future demand, from their point of view. This was put in terms of their customers telling them what they have in mind for the future and, on the basis of this, the firm itself thinking out what it wants to do. When asked whether the firm thought in terms of expanding the range of its customers, the executive said that it certainly did, that there has been an expansion in the roster of its customers. At the same time, it was pointed out that in planning for the future, members of the firm relied "pretty much" on the prospective planning horizons of their customers as they exist at some point in time, that is, on the nature of plans and the types of products that will be wanted. Implicit in the conversation was that they build into this projection some expected increase in the range of their market.

When asked whether they sought any advice or have any staff members that concentrate on the broader economic scene, the executive indicated that the firm takes advantage of an outside economic counseling service and that, at intervals, staff members from this service present to the Board a report of expected macro-economic conditions. This is supplemented by the written reports that form part of the service provided by the outside firm.

When asked about change in personnel and training of personnel, it appeared that the firm, like so many other companies, was very much aware of the fact that its personnel had been employed in the steel industry for most of their working lives, many with the particular company itself for most of their working lives. There seemed to be the feeling that for certain executive functions, particularly in finance, an individual's long history with the industry was as important as in production and sales, in which there seems to have been a pattern of promoting individuals who had long been associated with the industry, if not the company. The firm seemed very interested and anxious to recruit young college graduates, to get them into the firm early and start them along as a reservoir of future personnel.

This concludes the account of the four "representative" interviews. However, a few additional comments might be made. At some of the interviews, assistants, advisors, or other officers of the respective firms were present, and interspersed with the remarks reported above were comments by these individuals. One recurring theme involved financing as it affects decisions to expand. The assertion that the tax structure does not allow for adequate depreciation charges was stressed, i.e., that it puts a burden of financing or re-financing on the company that is more than it can bear. It was asserted that this has marked impact on the rate at which expansion is going to be undertaken. Yet we must note the recurring pattern of internal financing when these very firms do embark on new projects.

One over-all impression should be recorded. In interviews with officers of smaller or follower firms, it appeared that the individuals were much less braced for an "onslaught" and much more relaxed in their discussion. This, of course, may have been because of the specific personalities involved—they may have been more easy-going. But it does appear that in firms assuming the follower role, executives are less intensely public relations conscious—in the Madison Avenue sense—compared with the larger firms. It is difficult to know how far one can carry such a generalization, based as it is on such fleeting impressions, but there may well be some institutionally set difference that exists in an oligopolistic industry, between large and small firms (or the leader and the follower) which manifests itself in terms of differences in managerial personality or preoccupation. There was a noticeable difference in the demeanor of individuals representing the two types of organizations; those of the smaller firms conveyed a feeling of open frankness compared to a rather *studied* frankness in the other case. There was one other observable difference: those of the smaller firms conveyed a sense of more pure profit-maximizing entrepreneurship. They were quite aggressive in their assertions that they were actively competing with other larger firms in the industry and that they had every intention of going after potential business in particular market areas. At the same time, however, there was an acknowledgment that they did follow others as far as prices were concerned. This, therefore, left the listener with the impression that when competition was referred to, it was visualized as appearing in some form other than price competition.

In speaking to some of the staff personnel after the interviews, it was pointed out to me that a change in leadership has appeared over the past twenty years. In the case of some of the larger firms, observers from outside of the industry had in the past viewed a collection of separate empires, terribly sluggish, inefficient, and with a lack of dynamic qualities; they saw firms burdened not only with inefficient management but with inefficient production techniques.[74] When, however, these same firms had been examined more recently, the differences in personnel, demeanor, activity, the whole outlook of some of these firms, was remarkable; i.e., they had gone from positions of marked inefficiency to industry leadership in terms of efficiency. The implication was that the largest may indeed be the most efficient.

One factor, they explained, in creating this new efficiency in the case of the Steel Corporation, for example, was its employment of successful cost accounting or so-called standard accounting procedures that have been instituted throughout all phases of the firm's operations. The point was made that it was certainly to the credit of recent administrations that such changes had been brought about and that this was a really dramatic example of the effects of change in personnel and the importance of the perspectives of personnel. It was an indication that such a change could bring about profound changes in the performance of an establishment. They stressed that the ability to effect changes in efficiency was not linked to the size of the establishment and that, even in a larger organization, there is a freedom which allows choice, so to speak, as to whether to be efficient or not. Whether to use so-called progressive methods of administration, accounting, and production or not are decisions in which the size of the firm is irrelevant. The decision to use professional economists or consulting economic services was also singled out as a fairly recent innovation, one that has had implications for efficiency.

In these peripheral conversations, the view was expressed that one of the ways that the industry should cope with any new assault by an individual or government agency attempting to constrain or regulate the industry was to win over the intellectual community to the industry's point of view, to get it to

74. They were citing opinions of specialists in investment banking houses who had the responsibility for reporting on industry conditions.

change its attitude about large industries and large companies, and to change the image of large firms in the public mind—that they are not necessarily inefficient and that bigness is not necessarily bad. The fact that leadership in the industry is conscious of "the problem" and is attempting to reach this group is, of course, apparent in the recent public pronouncements of industry leaders in interviews, speeches, and testimony before government agencies. To amplify this point, we will turn briefly to some of the publicly recorded expressions of one of the industry's leading executives as a supplement to some of the attitudinal material presented above.

We have, in the person of Roger M. Blough, Chairman of the Board of United States Steel, an individual who may well serve as a prototype of a "new" and increasingly dominant figure among steel leaders in the industry. Mr. Blough's public expressions, for example, reveal a point of view about the American economic system that has nostalgic overtones of an earlier era. In an address before the Economic Club of New York, he said:

> We are caught in a web of mutual interest—the broad examination of that unmanageable, mysterious, hydra-headed common national adventure which we call economics. For whether we attempt to mold it, to tolerate it, or merely to recognize it, each one of us lives by economic precepts. We are co-adventurers on the road which can lead, we hope, to plenty and to the good life for all.[75]

75. R. M. Blough, "Learning to Multiply and Divide," an Address before the Economic Club of New York, January 15, 1957 (New York, U.S. Steel, 1957), p. 2. He continues: "Long years ago, in our valley, there was a little white one-room school house. . . . The smaller front desks were just right for little tots; and the last row could accommodate, very comfortably, a twelve-year-old boy day-dreaming behind a propped-up geography book. With all eight grades in one room, there was, of course, a certain amount of confusion and distraction; and yet—in this undoubtedly inadequate school room—knowledge somehow slowly filtered through, and we gained the rudiments of an education. And as part of that education, we learned what many had learned before: to multiply and to divide. . . . Today, in this land, we are—in a very realistic sense—continuing that early lesson in elementary arithmetic; for it seems to me that the heart of our economic life lies in our

Within this context, we find a commitment to a "characteristic" set of values: the American standard of living stems from American productivity and this results from incentives to save and invest; the unique American growth experience which differentiates it from experiences in other countries is not that "we are stronger, smarter, more industrious, more virtuous than our European ancestry . . ." rather that, "search as you will . . . I think you will find that there has been one major magnifying factor in this process of multiplication and that it is outstandingly American. And that factor is freedom." The predominance of the freedom value tied to freedom in enterprise, "freedom to venture our economic freedom . . . freedom of opportunity, freedom of choice, and freedom to enjoy the fruits of our labor as we please," mark it as the central and crucial variable.[76]

He believes that it is more important to be concerned with the adequate flow of loanable funds than the many efforts toward more equitable distribution of income in the society and, beyond this, that the inability of private capital formation to stem the tide of inflationary pressures is the primary problem for our contemporary economy. He links this to the need for revision of a tax structure which discourages capital accumulation and provides inadequate depreciation allowances. And all is ultimately tied to the need to prevent intervention in the sovereign sphere of private enterprise.

In an appearance before the Kefauver Committee in 1957, Mr. Blough commented on this issue in connection with questions concerning whether administered prices were responsible for inflationary pressures. He said:

> The impulse of governments to extend their powers even farther still persists today among a number of sincere, patriotic and well-meaning members of Con-

---

learning how better to multiply our economic capabilities, and to divide the values created by our multiplication among those who contribute to it. How well we perform that operation is, I think, a true measure of our total economic progress and of our rising worth as a productive people" (p. 3).

76. Ibid., pp. 4–6.

> gress. . . . You conscientious and over-worked gentlemen who are running this country have troubles enough of your own as it is, without taking on all of ours.[77]

The most revealing word in this statement is "ours." To Mr. Blough the sovereignty issue is a very real and pressing one and his emphasis on this issue involves not only the issue of sovereignty vis-à-vis government interference, but the potential interference from labor unions. This, of course, was demonstrated in the steel strike of 1959-60 where, despite claims that inflation was of primary importance, the issue of managerial sovereignty in the area of work practices seemed to pervade the negotiations.

It is probably on this issue of sovereignty that Mr. Blough most specifically focuses. He is not a man unaware of the "Keynesian" implications of the need for effective demand to sustain the output of industry, nor the implications of this for full employment and growth. Mr. Blough is not a man who wants to construct a world without trade unions or industry-wide collective bargaining, nor a world in which the role of government can be minimized in some Smithian sense. Rather, his thinking may be placed in some appropriate neo-Schumpeterian framework in which the policy-making executive is cast in the role of the venturesome, creative factor in the economy.

There is one issue in which public pronouncements might suggest some conflict in Mr. Blough's perspectives, but in fact do not—the area of foreign relations. There is no question that he is sympathetic to and relishes the ability of America and particularly the steel industry to shoulder some of the burden of international responsibility. For example:

> In two world conflicts at a time when we had about fifty per cent of the world's steel capacity, we were able to furnish the counterweight which swung the balance in favor of democracy. And since World War Two, we have shared over sixty billions of dol-

77. *The New York Times* (August 9, 1957), p. 22.

> lars worth of our wealth with our friends in many parts of the world. That is no mean accomplishment for the youngest of major powers.[78]

When he emphasized the incursions of foreign steel production on the domestic market during the steel strike, however, he used this to point a finger at labor rather than at our foreign competitors. He leveled a charge at the unions for increasing the wage differential that existed between American and foreign workers, thus charging them with the responsibility for hurting American industry and allowing for the prosperity of foreign producers because of their ability to deliver steel to our shores at lower costs. He did not suggest, as some business leaders have, that the problem was the result of our having unwisely put foreign producers on their feet after World War II; rather, his conception of the enlarging world demand for steel is in part involved in his feeling that America should participate in the economic development of other countries.

The steel executive today finds himself facing very serious problems as to the amount of "protection" he might be willing to accept or seek as a result of the threat of foreign competition to his industry.[79] How does Blough expect to meet this challenge? First, he places considerable confidence in the ability of the industry, through technological change, to improve the productivity of American steel and hence to permit it to compete effectively in changing world conditions. Having said this, he turns to what he characterizes as "a far greater challenge" and expresses it as follows:

> For however successful we may be in maintaining leadership in the field of scientific research, our

78. R. M. Blough, "Steel in Perspective," A Talk Before the Newcomen Society in North America, New York, December 4, 1958 (New York, United States Steel Corporation), p. 3.

79. In his 1958 Newcomen address, Blough stated: "May I add that while recognizing the value of such expedients as tariffs and quotas, increased tariff protection is not the ultimate or basic answer because free nations must depend upon trade. Trade is mutually beneficial. Moreover, if experience teaches us anything, then certainly we have learned that the power to erect tariff walls is not one in which America enjoys a monopoly" (ibid., p. 10).

> greatest challenge relates . . . to what we are going to do as individuals, as economic groups, and as a nation in the face of this fermenting world. . . . Can we in business and industry fully grasp the meaning and the implications of this ferment? Can we aid our fellow men to understand it so that they in turn may act in the common interest? And can we bring to our elected representatives in Government a real and dispassionate comprehension of the part they must play if they are to enhance—and not weaken—our nation's position among the nations of the world?[80]

And having posed what appear to be broader philosophical questions, his focus moves swiftly to a specific point. He goes on to say:

> We know, for example, that America's ability to compete successfully with these other nations in the years ahead will rest primarily upon what is known technically as 'the formation of capital'—or, in simpler language, upon our capacity to acquire and to pay for new, more efficient tools of production. . . . In any free capitalistic society, the only way to acquire the tools of production is through profit or the expectation of profit. But how many of our people really understand the nature and the function of profit. Not one, I suspect, in a thousand![81]

He concludes with reference to the responsibility of industry leaders:

> And if America does *not* survive that battle, upon whom will future historians lay the blame? . . . Will they say . . . that we of business and industry—we who *know* and understand the facts—were too busy, too indifferent, too impatient or too inarticulate to teach these facts in the simple, straightforward lan-

80. Ibid., pp. 12–13.
81. Ibid., pp. 12–14.

> guage that our fellow man could comprehend and believe? Will they record that we, who have been so successful in winning public acceptance of the *products* of industry, just simply didn't bother to try to gain public understanding of the *profits* of industry?[82]

Two dimensions are apparent in Blough's expressed opinions. The first is a commitment to a set of neo-classical values, and an image of a working economy within a modern technological context that draws on some nineteenth-century entrepreneurial *élan*. This is reflected in statement after statement. There is a distinction between Blough's thinking, and that of the totally committed organization man with his belief that it is the organizational entity that has the dynamic content. The difference between this point of view and Blough's consists in the ingredient of the individual, although operating through an organizational structure. It is the dual aspect of his view that sets it apart. It is on the individual that one must rely for dynamic forward momentum, and it is in the organizational context that the individual can realize his most creative potential. He states this quite explicitly at one point:

> To me the heart of the matter is not that we have certain management techniques which do produce excellent results but that within voluntary groups of free men lie the genius and the will to originate and perfect those management techniques.
>
> To me it is not that these corporate groups are the primary generative source of the capital formation of our land but that freely formed groups have, through their own internally directed efforts, the unrivaled capability for capital formation.
>
> To me the heart of the matter, therefore, is not that corporate groups are the indispensable providers of the physical elements of our national well being; the heart of the matter is that through these freely

82. Ibid., pp. 16–17.

> formed and constantly evolving organisms of production, generative forces of great originality rise far above the individual imagination of any of its members, enhancing the role of *every* man and giving breadth and scope even to him who may be called the uncommon man.[83]

In other words, for Blough, there is a corollary to speaking of the individual. There must be a working context, one which nevertheless leaves the individual's sovereignty intact, or else a completely permissive and voluntary environment.[84] This serves as a qualification to his confidence in the individual inevitably performing socially responsible functions—i.e., the laissez faire image. In his words:

> For one who has faith in the voluntary way to obtain production the answer is clear—it is the most useful means available and the means most likely to accomplish, both domestically and internationally, the end result desired by those who make our national policy. Its great characteristics are its flexibility and its capacity to utilize most effectively in production the combined talents of the members of

83. R. M. Blough, *Free Man and The Corporation* (New York, McGraw-Hill, 1959), pp. 39–40.

84. In a report of an interview with Mr. Blough after the 1959 steel strike had been under way for about two weeks, Mr. Blough issued a statement to the press, as follows: "Whatever the length of the strike, and whatever the eventual outcome of the negotiations—so long as they are voluntary—we in the United States Steel do not intend to raise the general level of our steel prices in the foreseeable future." The news items included the following: "But Roger Blough, chairman, said the company would hold to this position only 'in the absence of an involuntary settlement mandated by some public body or authority.' . . . Reporters pressed Mr. Blough as to what he would consider a 'voluntary' or an 'involuntary' settlement of the strike of 500,000 steel workers, and drew repeated implied warnings to the Government to stay out of the present negotiations. . . . When asked if the statement's phrase about 'so long as it (a settlement) is voluntary' could be rephrased as 'so long as we (the industry and the union) are left to ourselves,' Mr. Blough agreed, 'That's pretty close to what we had in mind.' " "U.S. Steel Pledges No Price Rise if 'Voluntary' Pact Is Signed to End Strike," *The Wall Street Journal,* (July 29, 1959).

> the individual productive groups which are the essence of our voluntary system. In the long run the unhampered voluntary approach has the inherent strength to prevail; only through such means can our national economic objectives as a people be achieved.[85]

Having thus presented statements which reflect the attitudes of leadership in the industry, we now turn to this same range of material to treat it in a somewhat more systematic fashion, and over a longer period of time. The following section is intended to be complementary to the foregoing.

CONTENT COVERAGE. The second body of evidence we shall examine concerns a fairly formal analysis of the content of pronouncements by leaders in the industry, i.e., a sampling of public statements made by them. Specifically excluded in this coverage will be statements of so-called professional economists writing about these issues—those who, in effect, can be considered secondary sources. Rather, the sampling covers statements cited in periodicals, in financial statements, and speeches before meetings of the American Iron and Steel Institute and other such meetings. The sampling is confined to a survey of the material where remarks were directed to the problem of expansion of capacity and the relationship of the individuals' work or company activity to the national economy, or perhaps something more remote, but indicative, such as "the general economic welfare of the country."

"Content analysis" is an analytical tool, developed largely in the field of political and propaganda analysis, and used by sociologists and psychologists to evaluate the substance of communications material in a way that allows some kind of objective reporting on the ideas, sentiments, motivations, or context found in the material. Formally, it has been defined thus:

85. Blough, *Free Man and the Corporation*, pp. 102–03.

> Content analysis is a research technique for the objective, systematic, and quantitative description of the manifest content of communication.[86]

Typically, this approach involves the actual counting of particular items, categories of ideas, intentions, or values found in items of communication or so-called units of communication. A tabulation of "responses" is then presented as an indication of the content or the trend in content of the material screened.

Taken in its most rigid form, this analytical technique involves a degree of "compulsiveness" in screening the material that is beyond what is necessary for us in this study. Therefore, our approach to the problem uses the essential format of content analysis without attempting to restrict the analysis to the tight rules of the game. It might be noted that the "creators" of content analysis, or at least some of them, are willing to acknowledge a kind of qualitative content analysis approach which is appropriate for interpretation, and which has far less rigid rules for sampling and coverage. One of the men identified with this field makes the point that for certain purposes—indeed, it appears, for purposes here—that a looser, more fluid approach to analyzing the content material is *more* appropriate. What is desired here, after all, is to disclose something about the motivation of these men and to use content as a key to something else, i.e., to a set of basic attitudes. Our hope is to gain

86. B. Berelson, *Content Analysis in Communication Research* (Glencoe, Illinois, The Free Press, 1952), p. 18. Three general assumptions forwarded with this definition are, as follows: "(1) Content analysis assumes that inference about the relationship between intent and content or between content and effect can validly be made, or actual relationships established." This important assumption justifies the use of content material to reflect purpose or motivation or attitudes and the technique will be drawn upon for our purposes in this study on this assumption. "(2) Content analysis assumes that study of the manifest content is meaningful." The presumption here is that both the individual expressing himself and the listeners and analysts draw on a common sense of meanings so that the communications sequence is not thwarted by semantic difficulties. "(3) Content analysis assumes that the quantitative description of communication content is meaningful" (pp. 18–20). The implication of this final assumption is, of course, that the relative frequency of theme or attitude reveals something about intensity and timing of expressed opinion. Further comment on this point will be made below.

insight into the attitudes or opinions of those making the statements screened.[87]

There is, however, one aspect of content analysis which can be deadening and lead to a lot of wasted energy; that is, when the investigator embarks on what perhaps should be called a fishing expedition, when he roams around in the public pronouncements of individuals he is interested in (or particular types of literature appropriate to his subject) made over some period of time, and comes up with whatever is found. This kind of open-ended searching would be useless and inappropriate here. Instead, I have attempted to identify a period of time which not only reflects a representative spread of opinion on the subject of investment programming in the steel industry, but one which (it could be assumed, a priori) reflects some kind of change in attitude. A series of pronouncements were then examined to see what the nature of the content was vis-à-vis our subject and whether the a priori known changes in attitude which occurred over the period were indeed reflected in the pronouncements of the individuals screened. It was in fact known, from change in behavior, that the period from 1947 to 1951 did reflect a change in the attitudes of the industry. The question to be answered was: would this appear in pronouncements from the industry? Therefore, the survey was confined to this period of years and to individuals who could be considered line officers within the industry.

Short of some involved psychoanalytic procedure, there is no assurance that public pronouncements do, in fact, reflect the true feelings of a speaker. But it must be emphasized that so far as the policy implications of the issues under examination are concerned, the public posture of leaders in the industry is of significance, if not of all-inclusive importance. As has been

87. As Berelson puts it: "The basic logic of this application of content analysis [to identify the intention and other characteristics of the communicators] is this: the content has such-and-such characteristics, therefore the communicators have such-and-such intentions. As in the case of cultural patterns, the relevant characteristics of the communicators are studied indirectly by this method only because the content is more readily available" (ibid., p. 72).

suggested at various other points, policy prescriptions, short of actual dictation to the leadership, will have to rely in large measure on what this leadership sees to be its performance of a social role and will have to rely in some measure, at least, on the desire of the individuals involved to conform to certain public pressures. It is relevant, therefore, to consider industry leaders' efforts to communicate to the public their values. While elements of rationalization and public relations awareness are inevitable components of public expressions, responsible leadership cannot so belie the actual sources of motivation and so ignore the set of stimuli which prompts action that they will nowhere appear or will appear only in completely distorted form in public pronouncements.

Though some may have reservations about this material on the ground that hidden inner feelings do not emerge, the true risk is more likely to be in using this kind of material to draw inferences without reference to other evidence; that is, in blind reliance upon this evidence alone. The reader is asked to temper his first reaction to the summary table to be presented below, because at first glance it appears to be one-dimensional evidence, based on a relatively limited sample. The analytical package below contains an evaluation of the results which links them to other empirical materials either presented here or elsewhere.[88]

88. Perhaps an explicit differentiation between what is typically referred to as quantitative analysis and what might be called a more qualitative analytical approach should be made. Berelson notes: " 'Qualitative' analysis is relatively less concerned with the content as such than with content as a 'reflection' of 'deeper' phenomena; this too is not an intrinsic difference between the two modes of analysis but it does seem to hold as an inductive generalization." He goes on to say: "In 'qualitative' analyses . . . the concern is more often centered on other events for which the content is only or largely a convenient indicator. Thus analysis of the agitators' propaganda is meant to reveal 'ideologies and ideological manifestations . . . the meaning of demogogy' and indeed to expose a whole sector of the society . . . psychologically oriented studies of content are primarily concerned with developing therefrom an insight into the personality of the creator—whether in projective tests of 'psychoanalyses' of the literary figures. In short, as compared to the quantitative analyst the interest of the 'qualitative' analyst lies less often in the content as such and more often in other areas to which the content is a cue, i.e., which it 'reflects' or 'expresses' or which is 'latent' in the manifest content" (ibid., pp. 123–24).

As to formal procedures, the approach calls for the designation of the unit of communication to be used, the subject matter categories, the orientation of the statements made, and finally the values reflected. The unit to be drawn on, as indicated above, will be the publicly expressed statements of leaders in the industry drawn from such sources as statements in the *Yearbook of the American Iron and Steel Institute,* quoted statements in newspapers and periodicals such as the *Commercial and Financial Chronicle,* the trade publication *Iron Age,* etc. The a priori "knowledge" of the observer is most crucially applied to the choice of topics that form the relevant subject matter categories. In other words, the material was screened and tallied in terms of coverage of a specific range of subject matter topics, 23 in number, as follows.

Condition of excess supply: an expression of the view that the industry was currently in or faced a situation in which supply of raw steel, semi-fabricated, or finished product was well in excess of the current or foreseeable demand for these products.

Excess inventory: the view that inventory accumulation was such as to exceed normal inventory proportions for the industry at some point in the production process, in some cases coupled with an inventory accumulation felt to be in excess of what foreseeable demand would justify.

Excess capacity: the view that the industry was experiencing capacity limits well in excess of probable demand, often coupled with references to historical experience concerning the characteristic excess capacity problems faced by the industry.

High wages: the assertion that the industry faced high wage conditions which (1) affect its ability to deliver the product at costs consistent with demand or with the division of returns to the factors, or (2) have implications concerning the relative ability of the industry to cope with competing products or competing producers in other countries.

High taxes: the assertion that with the tax structure as it stood, the industry was incapable of either adjusting its price structure to competitive levels or fulfilling its role as a capital formation agent.

High costs: the assertion that high costs in some generalized sense (or linked to references to high wages and/or high taxes) were detrimental to the industry's ability to compete and/or form new capital.

Low prices: another reference to the inability to form new capital as a result of inadequate revenues because the product of the industry was underpriced.

Low profits: a series of arguments focusing on profit-making as it affects the ability of the industry to form new capital (and at times linked to the wage-tax-cost picture).

Low future demand: reference to expectations of low demand in relation to the industry's ability to satisfy it and hence a reluctance to expand capacity.

Experience of the '20s and '30s: an explicit historical reference to the past under-utilization of capacity with the implication that attitudes regarding the future are conditioned by reference to the past.

Union strikes and slowdowns: an emphasis on the particular institutional setting that manifests itself in interruptions in production or in high wage–high cost-price pressure problems which ultimately impinge on the ability of the industry to finance expansion.

Raw material shortage: the belief that the limits of expansion for the industry will be conditioned by the industry's access to raw materials, particularly ore sources and, as a result, the possibility of this bottleneck factor inhibiting expansion planning.

Labor shortage: the fear that the inability to recruit and/or train labor adequate for industry needs will act as a constraint on the ability to expand capacity.

Optimum capacity: the view that the industry was currently at a level of capacity completely appropriate and adequate to the needs of the economy and that there was no need to presume that this situation would not persist.

Increase in future supply: the view that the industry would in fact increase the supply of output through plant expansion and increases in productivity which would guarantee the economy growing amounts of steel production.

Inordinate demand: the acknowledgment that demand conditions seem to be on the rise and industry awareness of this, but the characterization of this as of some inordinate or short-term nature.

Need for capital and higher productivity: the view that the industry did in fact need to work to expand capacity and that this put the industry under pressure to increase its productivity and also to gain access to finance capital which would permit growth.

Sufficient raw material: the view that the industry need not worry about future planning in terms of any constraint on raw materials, that these raw material supplies would be forthcoming.

Inventory shortage: the view that the industry was experiencing an inventory shortage, that current levels were below normal but an implication that this was a short-run condition.

Patriotism and national security: the view that, in disregard of more limited profit-maximizing motives, the industry would be willing to expand capacity at any time when its patriotic obligation was to do so, where the needs of national security required it.

High profit: the view that there were in prospect profit-making possibilities sufficiently attractive to encourage the industry to expand its capacity.

Future population increase: the view that the increase in population would demand greater output of the industry and that the industry might well plan for this larger population.

Future increase in demand: the view that there would be an increase in demand and that the industry should appropriately plan for it; that it would arise through either an increased per capita consumption of steel (combined with an expected population increase) and/or increased substitution of steel in the society, and/or the need to satisfy demand in markets heretofore not served.

It will be apparent to the reader that some of these subject matter topics are directly contradictory, and it is very unlikely that the antitheses will appear simultaneously in the same communication. Of course, what is desired in this analysis is not only a tallying of the appearance of various topics, but the expected trend of emphasis that occurs in various pronouncements over time as well. Indeed, as will be seen below, there is a trend pattern which reflects a change in the attitudes of leadership in the industry. What we have referred to above as orientation, then, may be seen to manifest itself in this presentation as a change in perspective reflected in the industry statements; that is, what might be referred to as a "pessimistic" or "negative" orientation regarding the industry's "ability" to expand capacity will be reflected in the appearance of particular topic items in particular statements. The orientation—the pro or con emphasis—is embodied in the spectrum of topic items tallied.

It is clear that this form of recording will submerge the varying senses of "indulgence" or "deprivation" expressed. It should be made clear, therefore, that while varying levels of intensity of feeling did appear in the several statements covered, it seemed that to try to attach any analytical significance, or to try to gauge intensity, would not only be hazardous, but beyond the needs of this study. At times, optimistic feelings were expressed in terms of (Lasswell's terms may be used here) realized gain, prospective gain, or avoided loss; feelings of deprivation or pessimistic attitudes were at times expressed as realized loss or threatened loss and, on other occasions, withheld gain.[89] Whatever the particular form or intensity, the tallying was done explicitly in terms of whichever subject-matter topic was given an important place in the statement screened.

As regards the final standard content analysis dimension, values, it would seem that the kind of material screened falls far short of being able to permit sufficient insight to make statements on these grounds and, therefore, "evaluation" will be treated among the summary remarks at the end of this chapter,

89. Berelson cites Lasswell's classification system (ibid., p. 151).

conclusions which may draw by implication on the content analysis, but not explicitly so.

Table 22 presents the tally sheet for the items screened. It will be noted that the column headings correspond to the 23 topic items referred to above. It will also be noted that each row, representing a particular public statement, is arranged in chronological order and that the coverage runs from the early '30s through 1952.

Several observations may be made regarding the distribution of entries in this table. As will be seen, a line has been drawn vertically to the right of the column headed Optimum Capacity. All of the topics entered in the column headings to the left of this line may be considered to be expressions of pessimism, or negative with regard to prospects of expanding capacity. To the right of this line are the more optimistic expressions; the optimum capacity category may be considered neutral. It should also be noted that the communications which lie above the horizontal line reflect that period preceding the Korean hostilities; those lying below it, the short period 1950–52 following the outbreak. If one were to generalize from this scatter, a downward sloping distribution appears which suggests a concentration of entries in the second quadrant (i.e., negative attitudes before Korea) and in the fourth quadrant (more positive attitudes after Korea). The scatter seen in this table may seem to be less than conclusive, but it does corroborate conclusions drawn from other empirical material. More than this, we find that in the pre-Korean period, the two columns which do contain a number of entries are those which reflect expressions of the ability of the industry to "catch up." The industry was saying that in the post-World War II period it had planned sufficient addition to capacity (modest as it may have been) to meet whatever demand might be expected and, as seen in the inordinate demand column, the contention was put forward that what demand did exist was abnormal, that it derived from pent-up post-World War II needs at a time when the industry was preparing to satisfy non-military demands, and this demand was a temporary post-war phenomenon.

There is virtually no reference to the more optimistic long-run appraisals which appear in the fourth quadrant of the table in the pre-Korean period, with the exception of one item in 1935 anticipating increased future demand and one item in 1947 anticipating an increased future population. This latter item, however, should be identified: it is a statement by W. Sykes in 1947, a reference to which appears elsewhere in this study (p. 111), and, examined in the light of other statements he had made, it is clear that its intention was to emphasize the adequacy of the industry's output in the face of increased future population. In 1947 Ernest Weir made reference to an inventory shortage problem, but once again this tends to have a short-run dimension. The other more frequently cited category was the need for capital financing and higher productivity in the industry, and here there are examples in statements by T. M. Girdler in 1938, by W. S. Tower in 1948, by Irving S. Olds in 1949, by W. S. Tower once again in 1949, and also by T. C. Drummond, writing in *Materials and Methods* in 1948. With these exceptions, conclusive concentration in subject matter categories is found on the "negative" categories in the second quadrant before the Korean outbreak and in the fourth quadrant afterward.

Looking at the third quadrant, negative attitudes expressed after Korea are confined to statements by Benjamin Fairless in 1950 deploring union strikes and slowdowns, by C. M. White in 1950 referring to the high cost-low profit-low future demand and to the strike problems of the industry, a later statement in 1951 by White covering some of the same issues; also a statement by R. F. Sentner concerning raw material shortages, a statement by F. K. McDaniel covering questions of excess inventory-low price-low profit relationships, and finally by W. S. Tower in 1952 touching on the high wage-union strike situation. But for the most part, once again, the emphasis is on the more positive values. Even in the cases of most of the spokesmen cited, there were continuing references to expected increase in future supply, and acknowledgment of high demand conditions even if they were only temporary.

This screening represents coverage of virtually all of the statements made in the *Yearbook of the Iron and Steel Institute* and in other representative trade publications concentrating on the period 1947–52, with exclusive reference to discussion of expansion problems. Therefore, in terms of the total number of references to the subject matter related to this, it perhaps is surprising that there were not more items in the coverage. The total number of topic references was 188; the number of relevant statements was 45. Because of the time sequence covered, there is quite naturally a predominance of sentiment registered in the negative quadrant.

It is to be expected that a screening through the past ten years would provide a substantial number of references to subject matter in the fourth quadrant. Along with this, more recent experiences involving attempts to obtain a revised tax structure, and labor difficulties, would create a concentration of entries in the appropriate columns. It might also be expected that the patriotism and national security issue would tend to recede as the interval from the Korean period grew. The need for finance capital and pressures for higher productivity would increase, particularly as the industry has recently become more conscious of the need to adapt itself to new types of production with the new space and nuclear research developments. It can be expected that the raw material shortage problem would not be emphasized so strongly as it was in the period following World War II; the industry has made extensive moves to assure itself supplies of raw materials in the more recent past. One theme that would not be reflected, because of the definition of subject matter topics in the statement tally, but nevertheless worth noting here as it is a factor implicit throughout this study, is that the theme of federal government control and the industry's distaste for these controls reasserted itself after the Korean War broke out and then tended to wane as time passed.

An examination of the annual reports of a representative number of firms in the industry over the years 1947–54 provides us with a footnote in support of the trend disclosed in the content analysis above. Though perhaps it is more typical for a "chairman's letter" to concentrate on the performance, past and expected, of the particular company, it is not unlikely for comments to appear which touch on the national economic situation. This is particularly true in the cases of United States Steel and Republic Steel, where statements are made covering the industry as a whole and only indirectly on company affairs. For example, in annual reports of Republic Steel some of the following ideas were expressed. In 1947:

> A theory has been . . . expressed that the present abnormal demand is . . . a normal demand. . . . We are convinced that expansions of these proportions are not warranted today. . . . The experience of the 1930's demonstrated . . . it was not capacity . . . which insured employment and prosperity but . . . the availability of customers and markets.[90]

In the 1948 report, the outlook for 1949 was said to be "confusing." Reference was made to what was called "pent-up" demand resulting from wartime depletion of industries. The claim was made that stocks were approaching normal. It was acknowledged that the use of marginal facilities at prices higher than normal did indicate a critical supply-demand imbalance and a tentative conclusion was that equilibrium would be reached in the later months of 1949.

In 1949, "All signs indicate a continuing strong demand in the years immediately ahead. . . ."[91] This is an indication of changed attitude. It is very unlikely that such a statement would have been made in 1947 or 1948, and the phrase "all signs" does appear to refer to more than inventory conditions. In the 1950 report, the perceptible change in tone appears not as a fleeting reference but permeates the whole report: "It is expected that 1951 will see a continued high level of industrial production."[92] If one compares the treatment of this issue in

90. *Annual Report,* Republic Steel Corporation, 1947, pp. 4–5.
91. *Annual Report,* Republic Steel Corporation, 1949, p. 8.
92. *Annual Report,* Republic Steel Corporation, 1950, p. 8.

the 1950 report, in which expansion is spoken of in sweeping terms, to that of the 1948 report in which expansion was indicated, but only along specific lines, one senses the difference in attitude. Examination of the 1951 report shows a discussion of capacity made only implicitly, i.e., in the context of announcing expansion plans. In the 1952 report, general confidence is expressed in the long-run nature of demand, and throughout the period 1953–56, the same general tone of optimism is found; as regards future demand, there seems to be no discussion of the ability to meet it; this seems to be beyond question.

The foregoing serves merely as a footnote to lend substantiation to materials screened in the content analysis discussion. It can be said that with considerable uniformity, industry annual reports reflect this trend in attitude over the period 1947–52. An exception might be the case of Inland Steel, which had announced in its 1948 report a 17 per cent increase in production facilities. It used this, as a matter of fact, as ammunition for confronting the administration, saying: "It is illogical, therefore, that construction of government . . . ill-advised and socialistic proposals be made at the taxpayers' expense."[93] Another exception can be found in Armco's Annual Report where there was early emphasis (1947) on the nation's strong demand and Armco's expansion program in connection with it. "To meet the nation's seemingly insatiable demand for steel products, Armco continued its expansion program—conceived and planned during the war period. . . ."[94] However, after referring to pent-up demand "rapidly becoming satisfied" in 1948,[95] it arrived in 1949 at the view that equilibrium had reappeared. "During the post-war period of unprecedented demand for steel and steel products, Armco prepared to meet the competition of a normal market certain to come."[96]

93. *Annual Report,* Inland Steel Company, 1948, p. 11.
94. *Annual Report,* Armco Steel Corporation, 1947, p. 6.
95. *Annual Report,* Armco Steel Corporation, 1948, p. 4.
96. *Annual Report,* Armco Steel Corporation, 1949, p. 16.

INDUSTRY ESTIMATES AND PLANS: EXPANSION AND NEW TECHNOLOGY[97]

It will be useful to present in the final part of this discussion on entrepreneurial attitudes some material which indicates where the changed attitude following the Korean War carried the industry, in terms of expansion and adoption of new technological methods. An examination of more recent plans and estimates for expanding industry capacity emphasizes the change that we have said to be marked and significant in the pre-1950 versus post-1950 period, and sets the stage for a more current appraisal of the attitudes and expectations that policy makers must face. The question to be discussed here, of course, is whether the shift in attitude documented above can be expected to be followed by similar shifts or whether this change was one of a fundamental and reasonably enduring nature.

The radical change in attitude toward expansion that followed the Korean period has been suggested. What one finds in looking at industry capacity figures for the '50s is a confirmation of this changed attitude, and the conversion into reality of the many plans that were put forward in the early '50s as expansion that was *expected* to take place. Further indications of changed outlook can be found in a McGraw-Hill study which shows the industry expected percentage increases in the physical volume of sales for the period of 1956–57 of 4 per cent and for 1957–60 of 16 per cent. This is confirmation of earlier assertions regarding the linkage between projected sales forecasts and expansion planning. And as regards capital spending planned in the steel industry for 1957, 60 per cent was for expansion proper, 40 per cent for replacement and modern-

97. As a supplement to discussion of more recent industry expansion estimates and plans and its adoption of new technology, reference to British experience provides an opportunity to examine a set of entrepreneurial decisions and attitudes in the context of a rapidly changing environment. The British case also focuses on changing locational criteria of significance to steel producers everywhere. Cf. my "Locational Decisions and Their Relation to Industrial Development," *Explorations in Entrepreneurial History, 3* (1950), 103–34.

ization. For the period 1958–60, the expansion figure drops to 53 per cent, with the balance for replacement and modernization, but we observe that a fairly consistent figure over-all remains.[98]

As a corollary to this expansion data, the McGraw-Hill survey also supports the contention of a changing attitude in indicating a trend in the industry to commit more scientific manpower to its productive efforts. Percentage increase in employment of scientists and engineers for the industry, estimated for 1956–57, is 5 per cent, and preliminary estimates for 1957–60 shoot up to a figure of 26 per cent.[99] For 1956–57, McGraw-Hill indicates an increase of 42 per cent for planned capital spending in the industry, though its projections over the period 1958 through 1960 do reflect a downward trend (1.3 billion dollars in 1958, 1.1 in 1959 and 1.0 billion dollars in 1960).[100]

Figures for steel capacity at the end of 1959[101] indicate a productive potential of 148,570,970 net tons (an increase of 937,000 net tons over the previous year). This marks an increase of steel-making capacity of 62 per cent over the figures at the end of World War II. Reports on the breakdown of this capacity indicate that the greatest jump was made in electric furnaces (14,395,940 tons a year as opposed to 13,495,138 at the beginning of 1959, or an amount accounting for 9.7 per cent of industry capacity). Capacity employing the recently developed basic oxygen process increased by 124,240 tons a year during 1959, bringing the capacity of this type to 4,157,000 tons, a remarkable upsurge in a new technological process. This not only suggests the potential for increased productivity in the industry by converting to new processes, but also reflects the

98. "Business Plans for New Plants and Equipment 1947–1960," Tenth Annual Survey prepared by the McGraw-Hill Department of Economics, (New York, McGraw-Hill, 1960), p. 5.

99. Ibid., p. 13 The Survey covers approximately 70 per cent of total employment in the iron and steel industry.

100. It would seem that more reliance can be placed on the validity of these expressed plans than perhaps in others sections of the economy, as the already noted longer gestation lag or planning lag holds here more firmly than elsewhere.

101. *The New York Times* (December 31, 1959), p. 25.

changed attitude toward seeking out and adopting radically new methods of production. It marks a fundamental characteristic of the post-Korean thinking of leadership in the industry.

The basic oxygen process is only one of several profound and significant changes to which the industry has turned its attention in recent years, and at least three of these are worthy of mention. The first is a series of efforts to develop new ore sources, such as the already mentioned pursuit of new ore sites in Labrador and Venezuela which gave rise to the rationale of locating newly integrated works on the East coast. The search for this ore was extensive and involved not only penetration into wild and unexplored areas of remote locales, but also the costly construction of transportation, warehousing, and docking facilities in these areas. The Venezuelan explorations date to 1947, when aerial mapping of Venezuelan territory south of the Orinoco River led the industry to an abundantly rich, hilly territory named Cerro Bolivar. Further exploration, drilling, testing, and so forth was necessary to verify the potential of approximately 500 million tons of ore, appreciably better than much in the United States and, in relative terms, reasonably accessible.

By the early '50s, United States Steel was in a position to expect shipment; a continuing flow from this source moved to their production facilities after the beginning of 1954. These finds meant the displacement of plans to press for ways of converting and preparing lower grade ores available in the United States for extensive use. Taconite ores could have been drawn upon, but would have involved large capital outlays to set up facilities to separate the impurities from lower grade ores. The alternative of relying on foreign sources was the more economically rational move, and United States Steel chose this course in Venezuela, Inland and Bethlehem in Canada. Taconite processing, though abandoned by some of the firms with access to foreign sources, still is considered a viable undertaking by firms without these sources and also for reserve capacity purposes. Republic Steel, for example, with a 50 per cent interest in Reserve Mining has indicated it intends to pursue the beneficiation of low grade ore in Minnesota by operating a huge taconite mining operation.

More radical than innovations to retrieve low grade ores are a series of recent developments in a process to reduce iron ore directly, eliminating the blast furnace stage of production. This so-called direct reduction process involves the converting of the ore to a dense metallic material which would lend itself to direct charging into steel furnaces. The underlying virtue of this approach, of course, is that the extensive capital costs of blast furnace facilities can be reduced if this process proves to be a useable one. Beyond this, experimentation has indicated that there will be fewer impurities in this type of charge than are found in pig iron produced under conventional methods. The direct reduction product can act as a substitute in open hearth charges for the scrap component of the charge which sometimes is unavailable or excessively expensive. Republic Steel and National Lead had, in 1958, invested approximately $5 million in a pilot plant in Birmingham, Alabama, which was capable of turning out approximately 50 tons of five-pound "iron briquettes daily."[102]

Actually the most advanced experimentation in this area has been done in Europe, where the drives to avoid the high cost of coking coal, scrap, and expansion costs of blast furnace capacity are more pressing problems. Among the virtues of the direct reduction process is the fact that it is not dependent upon coal as its only fuel source and can rely on lower grades of metallurgical coke for the carbon source necessary in steelmaking. Many steelmakers are dubious as to whether this process can be effectively applied to large-scale operations, but there seems little doubt that exploration in this area will be pushed, if for no other reason than the industry's consciousness of the spiralling capital costs of expansion of conventional facilities. News accounts describing one variation of this process, pioneered by Bethlehem Steel and Hydrocarbon Research, Inc., give some indication of its differing nature:

> To reduce iron in the H-iron process, heater ore is dumped into the top of a 90-foot pressurized tower on a kind of bed with a stream of hot hydrogen blow-

102. *The Wall Street Journal,* (October 16, 1958), p. 1.

> ing up from below. The ore moves down the tower in about six hours, cooking two hours on each of the three beds. The ore is kept at about 900 degrees Fahrenheit to make sure it doesn't melt and foul up the system. When the charge leaves the bottom reducing bed it is in the form of a metal sponge that is over 90 per cent pure iron.[103]

A ramification of this development, related to our earlier discussion concerning the proposal for a New England steel mill, is that there is the possibility, with capital costs in the H-iron process something less than half of what blast furnace and coke oven costs would be for the same capacity, that an integrated electric furnace and H-iron processing facility could be built for significantly smaller outlays than traditional integrated facilities require. Another significant feature of this development would be its ability to use a great range of differing grades of ore and, in particular, would allow the industry to sidestep the taconite refinement process. It is asserted that, in the Republic-National Lead process, ores with as low an iron content as 34 per cent and with as much silicone as 30 per cent can yield briquettes containing more than 90 per cent iron and something less than 3 per cent silicone. Moreover, this can be fed directly to steel furnaces.[104]

A second radical development on the horizon, with some similarities to those already mentioned, but with specific implications for the fabricating stage of the steel-making process, has been undertaken by Republic Steel. This is a method which not only circumvents the blast furnace and melting process, but involves a procedure in which "finely powdered ore with a high iron content is pressed into steel strips."[105] Commercial production using this process is not expected to materialize until the middle '60s, but again the capacity cost would be in the neighborhood of half of what similar facilities using conventional methods would cost. The process is intended to bypass the blast furnace, open hearth, coke oven, and blooming mill

103. Idem.
104. Ibid.
105. *The New York Times* (June 19, 1959), p. 32.

phases of conventional steel making and involves three major steps:

> First, iron ore is highly purified and reduced to metallic powder. Second, the powder is compressed into a semi-solid strip by rollers. Finally the strips pass through hot-strip rolling stands. The rolling stands reduce the strips to the desired gauge and roll them into coils which have the same quality as conventionally made steels. . . .[106]

There is indication that experimentation along these lines had been under way in Europe for some years, particularly in Sweden. But here again the development of the process has not reached the point where it can be handled commercially on a large scale. Nevertheless this, too, seems to hold promise for radical departures from conventional practices.

The so-called oxygen converter process is the post-war innovation which has materialized as an operationally feasible innovation. Although not as radical as the two previous examples, it is nevertheless of great significance. The process relies on the injection of a jet of high purity oxygen to the surface of molten pig iron and scrap, resulting in a profound thermo-chemical reaction that reduces the carbon content and converts the molten metal to high quality steel. The process shortens the refining period by an interval approaching eight hours—down to twenty-two minutes—and involves capital outlays of approximately $25 less per annual ingot ton of capacity than would be involved in the construction of conventional open hearth furnaces. At the end of 1957 Jones and Laughlin inaugurated oxygen converter facilities capable of producing approximately 750 thousand tons of ingots per year. Kaiser Steel has embarked on production using this method at Fontana, California, as has McLouth Steel in Michigan. U.S. Steel, Jones and Laughlin, Armco, and Pittsburgh either have a recent oxygen plant, or have made recent oxygen plant expansions, or are in the process of adding this type of facility to their existing sites.

In addition to capital savings, the process permits the use

106. Idem.

of more molten iron and less scrap in the open hearths, a problem referred to above. It also permits closer quality control. Basic oxygen capacity in the United States moved from 540 tons at the beginning of 1957 to a figure in excess of four million tons at the beginning of 1959, indicating a rise of nearly 650 per cent. Some predictions have been made that by 1965, as much as 35 per cent of the world's steel capacity would employ the oxygen method, and in the United States it would account for some 25 per cent of capacity.

As a final item which reflects the changed attitude and outlook of leadership in the industry in the period from 1950 forward, it is useful to examine figures developed by the Office of Business Economics and the Securities and Exchange Commission on expenditures on new plant equipment over the period under discussion. These indicate, for iron and steel, outlays of a fairly constant nature for the post-war years of 1947 through 1950, with a slight rise to 1948 and a subsequent decline from 1949 to 1950. There is then an appreciable jump beginning in 1951 and continuing up through 1953. A retrenchment, following the considerable outlays in 1953, follows in 1954, a slight upturn in 1955, and then a jump again in 1956–57 of proportions of the sort that occurred at the beginning of the decade; then slight retrenchment in 1958 once again. This can be seen in Table 23.

All of this evidence tends to confirm the attitudinal materials presented in the early sections of this chapter and suggests that change in perspective, revision of expectations regarding demand, and some more deep-seated and perhaps far-reaching revision of outlook with regard to the growth of the economy *did* manifest itself in selective expansion within specific firms. This, of course, has the ultimate effect of a general enlargement of the productive potential of this sector of the economy. A part of the picture, emphasized just above, has been adoption of new technology which may, in one sense, be considered to confirm the assertions by some that as new technology develops, the industry will leap to these new forms. On the other hand, an equally plausible alternative may be that accurate tracking of potential invention in earlier periods, accompanied by a lack of innovation on the part of industry, suggests that it is *a basic change in attitude* that explains the recent flow of innovation, rather than the fact that some great incipient technological revolution burst forth upon the industry in the period following the Korean War.

TABLE 23. *Expenditures on New Plant and Equipment, Seasonally Adjusted, for Primary Iron and Steel, 1947–58 (millions of dollars)*[a]

1947	633
1948	733
1949	608
1950	690
1951	1,175
1952	1,508
1953	1,253
1954	760
1955	853
1956	1,248
1957	1,725
1958	1,250

[a]Adapted from quarterly figures averaged to give annual figures of the U.S. Department of Commerce Office of Business Economics and Securities & Exchange Commission. (Note: unadjusted quarterly data by manufacturing industry may be found in *Survey of Current Business,* June 1956, and March 1958.)

There is substantial ground, it seems, not to turn one's back on the fact that a burst of new ideas did interact with a change in expectations regarding future demand, to bring forth the behavior just recounted. It is also quite plausible to see in the events of the period a basis for reappraising the competitive situation in the industry. The expansion undertaken by United States Steel for example on the Atlantic coast, or the efforts of companies to intrude in the Chicago market, are the results of revised managerial attitudes on the need for "competitive action." Surely the problem of facing depleted domestic ore reserves and the opportunity costs involved in beneficiation of taconite ores as opposed to searching out new ore sources has been important at this time.

There is another new element that appears on the more

recent post-World War II horizon that must not be neglected and which undoubtedly can explain certain aspects of the behavior of the industry as it occurred. This is the emergence of a changing international trade picture as regards movement of iron and steel products. There is no question that as the industry moved into the period 1957–59, the change in the net export position of the industry began to press on the leadership a new set of considerations which, heretofore, it had been able to ignore.

An article in the *Wall Street Journal* on this subject opens as follows:

> Italian-fabricated steel transmission towers will begin arriving in New York shortly for assembling along a 150-mile New York Power Authority line between Niagara Falls and Syracuse, New York. Akron's Goodyear Tire and Rubber Company and Firestone Tire and Rubber Company have turned to France and Belgium for certain types of wire for tire reinforcing. Borg-Warner Corporation of Chicago is experimenting with foreign steel, and finding the metal 'up to domestic quality.'[107]

Figures for 1957 on exports show an increase to somewhere near the mid-year point and then a continuing decline through 1957–58. In the third quarter of 1958 there seemed to be a slight revival, but from that point on a continuing decline occurred through 1959. On the import side, irregular movements of steel ranging from between 200 and something just under 100,000 tons of steel moved through to the second quarter of 1958, when steel import figures began to pick up and, though with some fluctuation, continued to gain, reaching in 1959 the neighborhood of 350 to 400,000 tons. In the last few months of 1958, the import figures rose above export figures and marked a new period as far as the foreign aspects of this problem were concerned; for a time, at least, the economy had passed over a threshold—the country had become a net importer of steel.

It is true that sales by foreign producers have been con-

107. *The Wall Street Journal* (July 29, 1959), p. 1.

centrated in a specific (and perhaps it would be fair to say a limited) part of the American market. Imports seem to be concentrated in product lines which take up only perhaps 20 per cent of the total United States market. These items are chiefly reinforcing bars, barbed wire, pipe, tubing, structural shapes, and wire products such as nails, staples, and fencing. Such items as cold rolled sheet, electrical sheets, tin plate, and other mass-produced, high-volume products are not typically part of the import picture at the present time. Of the items that are being imported, United States firms made wire products accounting for 5 per cent, pipe and tubing for 11.2 per cent, and reinforcing bars for 3.4 per cent of the domestic market.[108] One aspect of the import situation up to this point has been the greater logic of foreign producers shipping to the West Coast and Gulf Coast ports and perhaps to some East Coast ports. However, the development of the St. Lawrence Seaway—originally the hoped-for path for moving American products from the Middle West to the world—now seems to provide access for Midwestern manufacturers to foreign products at lower shipping costs.

It probably would not be too far from accurate to indicate that the situation, as it stands today, is one of fundamental disequilibrium. Patterns of trade for the steel industry are shifting, and it is perhaps too soon to see how permanent or how profound these changes are. It is not too soon, however, to record the impact that this new post-war development has had on the thinking of the leadership of the American industry. The editor-in-chief of *Iron Age* cryptically described the situation thus: "Foreign steel is now the realistic nightmare of the American steel maker."[109] He goes on to quote the president of Republic Steel, Thomas F. Patton: "First the foreign manufacturers took our foreign market. Then they went after our coastal markets. Now they're invading our inland market. Everyone in the industry feels that foreign steel is a growing menace."[110] Foreign

108. "Steel Import Surge: It Reaches Record, Seems Likely to Stay High," *Wall Street Journal* (July 29, 1959), p. 1.
109. *Time* (July 20, 1959), p. 94.
110. Idem.

competition, then, and revised attitudes toward secular demand for steel arising in the economy, as well as consciousness of new technological possibilities, combine to mark the changing attitudinal context of leadership in the industry today.

THE "REPRESENTATIVE ENTREPRENEUR"

When Marshall wrote of his "representative firm," he formulated a concept which allowed him to generalize about conditions in an industry and at the same time to acknowledge that individual establishments would depart from this generalization at any given point in time. Can a similar characterization be given to the leadership in this industry? Can we speak of a "representative entrepreneur?"

The answer would have been "No" some decades ago, but since World War II there does appear to be some convergence of personality type, a general set of attitudes, and a pattern of behavior. There are fewer of the heroic figures that dominated the picture even as late as the 1930s. In the 1960s, the job of summarizing industry attitudes is still risky, but a profile does emerge. It is possible to attempt an outline today because on the whole there is much more homogeneity in the leadership, though there are still powerful and distinct individuals within the industry.[111] Subject to "random variations," the material presented in this chapter provides us with some ideas about

111. This raises the question of the individual and the organization. It can be seen that the particular "ambitions" of the organization may be either conditioned by the history of the organization or affected by the force of a single or relatively few strong personalities in leadership positions in the organization—that their ambitions will determine whether a particular decision-making sequence is "right." For example, the degree of vertical integration decided upon and the market power or ambitions for market power of a given producer will determine what share of the demand facing the firm it will want to respond to, or to what increases in aggregate demand it will respond. It also must be recognized that strong organizational loyalties (as a yardstick for the individual decision) may have effects which are less than desirable from the point of view of over-all economic and social needs, i.e., the individual, so keyed to the organization, may well make decisions which are "correct" from the point of view of the organization and its policy, but which are recognizably at odds with values and needs in areas of concern outside of the organizational setting. This is, of course, the recurring theme of this volume and it is a problem to which attention will again be turned in Chapter 6.

decisions in the industry, and the men who make them.

Let us summarize the present behavior patterns and attitudes: in some respects, issues worrying industry executives are more disparate in emphasis today than in the immediate post-World War II period. There are also new concerns; for example, growth in foreign competition and emerging competition from substitute materials. Two notable changes are apparent and, because of emphasis, almost appear to be new items: first, the industry is markedly more conscious of changing technology, what it promises, and the necessity of adopting it; second, it is more deeply and subtly conscious of the need to "project" an image of socially responsible behavior. But some sources of concern remain unchanged: labor troubles, inadequate profit margins "needed to finance expansion," and future economic stability and growth. Over-all, there is still a long-standing recurrent bearishness about the future.

In weighing these issues and attaching some priority to them, industry leaders, despite new perspectives and constraints, still engage in maximizing behavior and still exhibit a primary interest in *profitability*. The modification that emerges tends to be more a reflection of the new personality type that heads these companies than a change in the fundamental production or market structure conditions in the industry.[112] Having observed this, however, we should, at the same time, recognize that there is a limit to the reshaping of a given personality to the requirements of the environment, and equally a limit to the personality's ability to avoid the requirements and pressures exerted by

112. In this connection, in her study of the 12 steel companies, Schroeder concludes: "A primary explanation for the fact that a firm has developed in a particular way is to be found in the nature of the entrepreneurship associated with that firm. The really significant entrepreneurial role was that played by the person or group of persons who put a given company 'on the map' so to speak, who shaped its pattern of development and established its basic policies. Generally, but not always, this central direction has been exercised by the man who succeeded in becoming president of a company or chairman of its board of directors. The vital roles exercised by Gary of United States Steel, Schwab of Bethlehem, the Blocks of Inland, and Verity of Armco have been pointed out, as well as the lack of such vital leadership in such firms as Crucible, Pittsburgh and Wheeling" (Schroeder, *Growth of Major Steel Companies,* pp. 209–10).

that environment. In all likelihood an element of natural selection controls.[113]

What specific qualitative picture then emerges from studying steel industry executives? For the most part, these men see themselves as performing in the Schumpeterian image—exercising their imagination, vigor, and boldness of spirit to push frontiers forward, to seize new opportunities, to engage in the competitive struggle with relish. Yet, they see this taking place within the organization.[114] They see no paradox attached to a dynamic role and the fact that they are involved in large bureaucratic structures and concentrated markets. This might be explained by the fact that the men studied were so high in the hierarchy that they were oblivious to organizational constraints. More likely, they genuinely believe the organization and the oligopolistic market place to be dynamic. Therefore, as regards *perception* of role and environment, we see grounds for assuming that a dynamic element is present.[115]

113. As one psychologist points out, in a study of business executives, though the role accepted and assumed by the business executive is subject to the requirements put upon it by the business organization itself and by pressure from society as well, the assumption of this role is a function of the personality of the individual as well as of these "externals." He goes on to note that though the "personality structure is reshaped to be in harmony with the social role," there are limits to this process and in fact it is recruitment which cuts down the "time involved in teaching the appropriate behavior. Persons whose personality structure is more readily adaptable to this particular role tend to be selected, whereas those whose personality is not already partially akin are rejected" (W. E. Henry, "The Business Executive: The Psychodynamics of a Social Role," *The American Journal of Sociology, 54,* [1949], 286–91).

114. When executives describe how decisions are arrived at, they usually refer to organizational structure in terms of communication patterns in the firm and the delegation of decision-making functions. The picture of "the team" emerges strongly; it is asserted that decisions are usually handled in a decentralized way and that then they are integrated into a consistent over-all policy for the organization. The idea of "group strategy" is stressed. As to form, the committee system is typical. Benjamin Fairless stated: "A lot depends on the subject matter, but as a general rule, I believe in group decisions rather than individual ones, especially when you are going to spend a lot of money" (J. MacDonald, "How Executives Make Decisions," in *The Executive Life,* by the Editors of *Fortune,* [New York, Doubleday, 1956], p. 174).

115. A question intensely debated and yet unresolved: Does the appearance of a socially conscious executive result in the submergence of economic man to the detriment of capitalism? These issues and the associated literature are reviewed in Richard Eells' *The Meaning of Modern Business* (Columbia University Press, New York, 1960).

The picture is not so clear cut, however. These men see their role as that of manager, not proprietor, and this requires an adjustment in the Schumpeterian image. These men feel a *managerial* obligation to stockholders, and though they insist this means *dynamic* management, it is management nonetheless. Different attitudes toward risk-taking and "preference-function" maximization inevitably follow. With this, and perhaps because of this, there is a predilection for standing on some middle-ground as regards estimates for the future. Though in their view they are pushing on to new frontiers, and recently very conscious of the implications of technological change, these men in fact tend to choose targets that fall short of the outer limits of innovational possibilities. They fulfill a leadership function, but one colored by a good measure of risk-aversion. In short, there is, for all the talk of desire to meet the needs of the expanding economy, a conservatism which permeates the performance of the role and the decisions made.

In addition, though steel executives see themselves as fulfilling roles of dynamic profit-maximizing leadership, they appear markedly defensive in some of their attitudes. Some of the statements concerning the secrecy of their estimates suggest that as much as there is fear of disclosure which would benefit labor or competitors, there is greater fear of giving government ammunition which would result in pressure on the industry. References were made in interviews to figures in government who were out to destroy the industry. Mention of friendly and unfriendly academicians were followed by assertions that this group had to be "won over." But more significant perhaps than these manifestations was a recurrent attitude of self-righteousness expressed in such terms as: "If the industry cannot satisfy *its* needs, capitalism as a system will suffer."

On investment decisions, there is a pattern of thinking in very short-run or very long-run terms—the former on an ad hoc basis, the latter in a generalized and quite different context. No model of the economy seems to be a part of the decision process. Set against the development of reasonably powerful analytical tools in professional economics, the businessmen's ideas are indeed cast in very simple terms.[116]

Considerable emphasis is placed on competition. Its presence is proved by the effort needed to maintain one's share of the market. Competition is thought of in these terms; prices are not mentioned. Moreover, competition is said to be reflected in the diversity of forecasts and the differences in the way various executives meet problems. It would appear, therefore, that these men have a near-classical conception of pricing in their markets, where demand is completely elastic and profit-making opportunities arise from their ability to lower or change the position of their cost curves.

A great deal of talk concerns problems of financing expansion. It is hard to separate and weigh the strands of thought which form this emphasis. Is it reluctance to acknowledge the possibility of internal financing? Is it drum-fire to bring about revised tax laws? Is it an attempt to set the stage for greater earnings ratios, to explain the level of expansion undertaken or foregone, or to forestall further wage demand or price increases? Whatever the reason, this is the area in which the industry acknowledges government can help and should help. Accelerated amortization programs, and more recently tax credits, are looked upon as legitimate, desirable moves by government.

Two final themes must be reviewed: bigness and executive leadership. The assertion that "bigness is not inefficient" is heard often, perhaps defensively. On the contrary, they assert that great size is the only way to meet the challenge of our expanding needs and thus in turn to protect our way of life. Steel must be produced on a large scale to permit economies, to keep apace of new technology, to provide employment, and to meet

116. In discussing his attitude towards the future, Avery Adams of Pittsburgh Steel observed: "I was born an optimist." And Joseph Block of Inland: "In the main, I am an optimist about the nation's economy and my own business." As to attitudes on the outcome of decisions some recognize the element of probability, e.g. Adams: "It is difficult to read the future in terms of the demand for steel. Estimates of steel executives have been on the low side every year for the past twelve years" (MacDonald, "How Excutives Make Decisions," p. 171).

the challenges from abroad. A large, efficient steel industry can do these things and fulfill this responsibility if it has the freedom to coordinate planning, achieve balance in its product-mix, and maintain flexibility. This must be accomplished within the present market framework and with existing personnel.

Expertise is the final point emphasized. The government personnel who have turned attention to the industry are generally considered inferior, limited in experience and perspective, and in some cases directly hostile to the industry. Great pride is taken in the industry's internal recruitment patterns of the past.[117] In fact, we observe considerable complacency within the industry as to the competence of management. This is seen not only in internal staffing but also in the willingness to forego aid in such matters as forecasting from professional sources outside of the industry.

In examining this point, we note that the successful administrative structure does rely on the establishment of special areas of concern for individuals in the organization, and does rely heavily on individuals developing specialized skills once the performance of assigned duties is undertaken. It is in fact this

117. It seems likely that the rate of mobility within the industry from one firm to another on the top executive level will remain unchanged. Characteristic of the industry is homogeneity of production practices, technology, and procedures. Compared with many other industries, secrecy and refinements of exclusive know-how do not exist in the extreme and hence the barriers to movement from firm to firm are not protected with the vigor that they are in other industries. It is not unlikely, as a part of this pattern, that the assumption of top executive positions in the industry will take place at a somewhat earlier age as men are wooed from one firm to another in the hope of strengthening the executive structure of a given firm. With the trend toward more preoccupation with the problems of public policy within a given firm and particularly vis-à-vis government relations, one might expect that executive personnel with law backgrounds will be recruited more and more in the future. Though this is a risky assumption, the future top executives may well be chosen not only on the basis of intellect, but in terms of "public demeanor," and this set of qualities, of course, may be backed by a variety of educational experiences. A little also can be said about the family background of the future leaders of the industry: as educational opportunities become more dispersed within the society, generalizations about homogeneous family background will prove to be less valid. For an empirical study of "Family Background, Training and Experience of the Top Officials of the Largest Corporations in the United States," with comparisons for the years 1899, 1923, and 1948, see M. Newcomer, *The Big Business Executive: The Factors That Make Him* (New York, Columbia University Press, 1955).

aspect of behavior within the organization that cultivates and leads to expertise.[118] Even given this distinction between generalized background and specialized performance of function, however, a surprising lack of explicitness is found when individuals in the industry are asked for a characterization of the decision-making apparatus. There appears to be an inability on the part of decision-makers to analyze the act. Consider, for example, a statement by Benjamin Fairless: "You don't know how you do it; you just do it."[119] Or, when another steel executive was pressed to indicate how a rate of expansion was decided upon, he said: "We have a feel for the right rate." There is undoubtedly an element of intuition in decision-making and, as we have pointed out in the first part of this section,[120] this is particularly true of so-called strategic decisions in which the planning horizon must be well beyond the time limit where extrapolation of trends and other manipulation of data is a reasonable exercise. Perhaps it does mean something to characterize certain aspects of decision-making as an "art." This may suggest elements of irrationality, but there are elements in a decision which draw on qualities of intelligence, past experience, and rules of thumb combined with qualities of imagination, venturesomeness and a feeling for a situation which may

118. As Simon has indicated: "to gain the advantages of expertise in decision-making, the responsibility for decisions must be so allocated that all decisions requiring a particular skill can be made by persons possessing that skill" (H. A. Simon, *Administrative Behavior: A Study of Decision-Making Processes in Administrative Organization,* New York, Macmillan, 1940, p. 10). Cf. also E. W. Bakke, *The Fusion Process* (New Haven, Labor and Management Center, Yale University, 1953).

119. Cited by MacDonald, "How Executives Make Decisions," p. 165.

120. Eisner also states: "When asked by the interviewer to comment on the inability of top management to explain pay-off calculations supposedly used in their companies, he indicated that top personnel did not get into the details of calculations; as long as they had a consistent measure they did not check further. [Footnote] in partial response to further inquiry as to methods of calculation, the company president wrote: 'It seems to me that you may be making too great an effort to blueprint the processes of determining capital expenditures. If they could always be done according to a specific formula, there would be very little need for individual judgment. However, in my opinion, once the facts are assembled, judgment as to the long-run value of the contemplated expenditure, all angles considered, is the determining factor' " (Eisner, *Determinants of Capital Expenditures,* p. 57 and note 32).

be labeled *expertise* and which, perhaps, should be included in a list of the components of the rational decision. This leads us back to the point made by executives on "advice" from outside the industry. The question remains whether the assignment of personnel from government to the ranks of, or to supervision of, private industry woud be a viable undertaking (even though certain conditions indicate need for national supervision of facilities) precisely because of the validity of the industry complaint, i.e., the lack of expertise possessed by re-assigned government administrators to industry positions.

In sum, executive attitudes do reflect some changes. They appear more expansionist in their thinking; the industry leaders do appear more "socially conscious"; they are generally committed to an internationalist position and are favorably disposed toward lowering tariff barriers. They have not modified, however, their hostility to any government agency turning its attention to problems within the industry; nor is there any noticeable introspection concerning the industry's market structure and its problems. In making these observations, we are not, for the moment at least, evaluating which attitudes, perceived roles, or view of their environment are right, wrong, desirable, or undesirable. Note that "socially conscious" leadership *may* be a drain on industry vitality—or that the internationalist position *may* have detrimental effects on wage policy. Similarly, firmness in maintaining a government hands-off policy—or maintaining the market structure as it is—may be just what is necessary in long-run social policy. We have attempted, up to this point, a characterization of practices, policies, and attitudes within the industry. We now must face the reconciliation of this information with the hypothesized macro-economic problems that may confront the national economy because of behavior in the industry. In the foregoing, we have moved through descriptive and analytical material. The final question is: what, if anything, should society do to modify this picture?

SIX NATIONAL ECONOMIC POLICY AND THE INDUSTRY

A GENERAL POLICY CONTEXT

Before proceeding to a discussion of specific policy questions, it will be useful to think in terms of an "ideal" picture, one that is not inconsistent with the *real* environment of the industry nor with national economic conditions and thought, yet is consistent with some set of long-run social, political, and economic goals. As regards the latter, present-day opinion suggests that we are anxious to preserve the context of "individual enterprise," modified to the scheme of entrepreneurial-manager-owner interaction which has evolved in recent decades.[1] In addition we see as a goal the cultivation of our potential for economic growth, but without doing violence to the pattern of distributive shares which is part of the capitalistic form of economic organization and which affects the broad cultural configuration we know as contemporary American society. Then there are behavioral considerations: we do not wish to disrupt motivational patterns which provide the momentum for production (and consumption) in the economy; at the same time, we wish to sustain political equilibrium. Although conscious of the need to prevent consolidation of power in the hands of fewer and fewer members of the society, we desire to sustain the motivational dynamic crucial to the operation of a privately dominated economy. In the absence of this sustenance, positively maintained, decision vacuums would require centralized administration of the econ-

1. Cf. pp. 160 ff. above.

omy and with this the transfer to what is an essentially different form of political and social organization.

Having said this, it would be well now to explore questions of the "ideal" functioning of the steel industry. We look for sufficient output produced at a point in time and, beyond this, output that is increased as the needs of the society for it expand, at a rate and in forms (final goods) that the society (rather than the industry) chooses. This suggests that not only is the industry willing to grow in some general sense, but also that a communication network (e.g., the price mechanism) is able to indicate the appropriate rate and product mix and that the industry is perforce sensitive to these communications, i.e., (1) they read the messages and (2) they respond to them. Finally, it is expected that the distribution of power and behavior constraints within the industry are such that so-called "undesirable practices" vis-à-vis competitors, customers, and society at large are absent.

Now as to some modification of these ideal formulations: the phrase "individual enterprise" must be altered to "corporate enterprise" or perhaps to "organizational enterprise," with the implication that structures and formulations must take into account the social-psychological organism that has taken the center of the stage today.

Of course, when reference is made to "growth potential," there is sufficient ambiguity to leave the question completely open. Growth for what? Is the significant factor the determination of some certain rate? Is it not that such a matter is so subject to exogenous factors, particularly military preparedness, that long-run speculation is quixotic? For our purposes, these questions may be reduced to the following: Is there a significant role played by a specific capital goods industry, such as iron and steel, which may affect growth, at whatever rate, subject to whatever influences? This has been discussed above: we observed that in short-run terms its potential as a bottleneck may be significant, and hence may participate in delays to secular advance, and in the long run, recalcitrance on the part of the industry to meet the demands of the society may force either a shift to foreign markets as a source of supply[2] or to alternative products as substitutes.[3]

The time-setting of a good part of our discussion has been confined to the short run by the very nature of the so-called bottleneck phenomenon. There is broader theoretical significance to this observation than the mere stressing that bottlenecks are a manifestation of short-run maladjustment. To have suggested that a bottleneck may persist into the long run would have been to describe a situation in which the economy never does come around, i.e., never does adjust[4] to specific scarcities.

2. However, it seems likely that foreign producers' advantages, if they emerge, will be confined to certain product lines. Cf. Louis Lister, *Europe's Coal and Steel Community: An Experiment in Economic Union* (New York, Twentieth Century Fund, 1961). One aspect of the need for growing levels of output concerns our competitive struggle with the Soviet Union. This argument places the position of the United States relative to the Soviet's as the key determinant in how fast expansion should proceed. Cf. John B. Parrish, "Iron and Steel in the Balance of World Power," *The Journal of Political Economy, 64* (1956), 382.

3. "Although competing materials have had only a slight impact on the steel industry as a whole, they have had marked effect in certain product lines. Competition by aluminum has been principally in light, flat-rolled products, including roofing and siding and containers. The increased use of aluminum in automobile motor blocks has been at the expense of cast-iron foundries rather than steel mills. The increased use of prestressed concrete in construction has cut into the market for structural steel. Plastics have become an important competitor for pipe and tubing markets to the extent that several steel companies are now producing plastic tubular products. Plastics are also used in a number of products formerly made of steel, such as toys. Other competing materials include glass, asbestos-cement pipe, and adhesives.

"Although new uses for steel have not been on a mass tonnage basis such as followed the invention of the automobile, and many have not been entirely new uses but rather familiar uses in new applications, nevertheless steel has in recent years come to be utilized in some areas for the first time. In the construction industry, for example, steel curtain walls have replaced other materials in the facing of office buildings. Air conditioning in many types of structures has provided a new market for steel, as has also the increased demand for such items as freezers, automatic laundry equipment, and dishwashers. In the military field, new, high-temperature alloys have been developed for use in rockets, missiles and jet aircraft, and similar products employed in nuclear energy equipment are finding increased use for nonmilitary purposes" (U.S. Department of Commerce, *The U.S. Industrial Outlook for 1961* [Washington, Government Printing Office, 1961], pp. 236–37).

4. A distinction is made here between adjusting to, and reconciling itself to, a scarcity: the latter may not accompany the physical allocative adjustment to a specific scarcity; i.e., the economy may never be able to satisfy certain specific wants.

This is conceivable, in a sense, where the distinction between long and short run is based, not on some type of adjustment criteria, but rather on some arbitrary passage of time. Who is to say what is the long run? But this is a definitional problem wherein the time interval may not be significantly long for the "long run" in the sense it has been used in our discussion. It is also possible to think of a persistent bottleneck where a disequilibrating factor is of a chronic nature and is of such integral and basic significance to the fundamental sequence of relationships that its scarcity blocks the flow of income payments or goods and services—for example, a chronic shortage of money or labor. But these would seem to be more appropriately designated as basic shortages rather than specific scarcities, in the sense that we have spoken of inadequate steel-producing capacity. It is also conceivable that once a specific shortage appears, the economy is of such a static nature that regardless of the passage of time, it is incapable of altering any relationships within its structure to eliminate the shortage. However, this hypothesis may be dismissed on the grounds that it is not representative of modern relationships in the American economy.

It seems reasonable, then, for us to assert that a specific scarcity cannot exist in the long run. We may emphasize the essentially transitional nature of the phenomenon which, in turn, suggests that policy formation must be focused on particular time intervals. The bottleneck and the potentially adverse effects that may accompany it are likely to be villains of the moment, something to be reckoned with in the immediate future, and a by-product of discontinuities in the course of change. If development proceeds "in balance," all will be well; and it may be assumed that in the long run, the specific problem will vanish, either through relief of the scarcity or through adjustment via substitution in the production process or in products demanded. There do remain, however, the possible problems created by the appearance of a bottleneck in the form of the possible choke-off sequence described above. And in addition, if the economy is passing through certain phases of structural change when steel is in short supply, the bottleneck formulation may have even more to say. Indeed it may be advisable to think of the phenomenon as operating in less precise terms than the word "bottleneck" brings to mind.

Duesenberry, in commenting on Hicks' model, has taken the position that a full capacity ceiling in capital goods industries is unlikely, because they too are subject to the same stimulus to expand as are industries in the consumer goods sector. However, as a specific example, he does cite the steel industry in acknowledging that a bottleneck capacity could arise where there was a time lag in the output of investment goods, so that, "If output were increasing at an increasing rate, output might catch up with capacity."[5] But then he rejects this as unlikely, because of the standby capacity characteristic of the industry. Gestation lags, however, might well be responsible for a specific scarcity in steel capacity at some point in the expansion phase of the cycle and while the feedback of rising demand in other sectors will act to encourage plant expansion in steel, the possibility also exists that management in the industry will be reluctant to expand in spite of the expanding market.[6] For example, the fact that steel, currently commanding a high price in the market, would have to be diverted to provide for increased steel capacity, might strike the entrepreneur as a self-defeating operation in that he might consider the very increase in output a threat to sustaining the price level. The question of entry and the relative nearness to an oligopoly structure in the industry would have to be considered here.[7] This reaction may stem from generalization of past experience or be based solely on an appraisal of future expectations. Whatever is the basis, the net effect may be to create the type of specific scarcity that has been under discussion.[8] For purposes of formulating policies with the

5. Duesenberry, "Hicks on the Trade Cycle," pp. 469–70.

6. For additional conclusions relating investment practice on the part of steel management to investment theory, cf. Andrews and Brunner, *Capital Development in Steel,* pp. 357–60.

7. For a discussion of the effect of shifting demand on oligopoly investment policy, cf. C. Kaysen, "A Dynamic Aspect of the Monopoly Problem," *Review of Economics and Statistics, 31* (1949), 109–13.

8. An additional possibility, not developed here, is that where a strong and persistent desire to invest exists in the economy, the scarcity condition may well tend to prolong the upswing.

intention of modifying cyclical fluctuations, therefore, consideration of entrepreneurial response patterns must be added to those presented in the foregoing.

The larger issue of the relative merits of the constrained cycle explanation of fluctuations remains, as does a question not yet discussed: what is the likelihood that the limit to steel capacity will be reached? Even if it could be uncontestably established that excesses of capacity, rather than shortages, were the norm as the upper turning point is reached,[9] the *possibility*—as in Duesenberry's example for steel—would still remain. Therefore, it is to the second question, that of steel capacity utilization, that attention will be turned.[10]

Capacity figures over the 40-year period 1900–40, show only five years in which production as a percent of capacity exceeded 85 per cent. Those were the years of World War I (1916–17–18), and at the peak of the boom in 1929; (also 1906, with a figure of 85.4 per cent). In the period after 1940, however, these figures rose to well above 90 per cent, and above 100 per cent (rated capacity) at some times. In the 1950s utilization figures drop once more. It does appear, at this writing, that this cyclical pattern will persist. Some qualification, however, of the generally consistent under-full capacity mode of operations should be made.[11]

Standby equipment is a characteristic part of the steel industry. As technical improvements are introduced and as equip-

9. E.g., S. C. Tsiang, who suggests that extensive capital development (inflationary) bucks intensive capital development (deflationary) and that the latter ultimately comes to the fore, yielding an excess capacity situation, "Accelerator," p. 341. Also see G. Haberler, *Prosperity and Depression,* pp. 374–75, A. H. Hansen, *Business Cycles and National Income* (New York, Norton, 1951), p. 484, and J. Steindl, *Maturity and Stagnation in American Capitalism* (Oxford, Blackwell, 1952), pp. 127 ff. The converse of the bottleneck condition, the glut, is not too significant a consideration here because of the practice of the industry to cut back production in relatively immediate response to changes in the market picture.

10. As is indicated in earlier sections, we contend that excesses of capacity are generally part of an over-all unemployment situation—from inadequate levels of national economic activity rather than excess investment in plant within steel.

11. Presentation of capacity-output relationships and investment for iron and steel is a part of Chenery's article, "Overcapacity and the Acceleration Principle," pp. 19 ff.

ment is replaced, the latter is often retained because of continued usefulness in handling peak load situations. One relevant question, in terms of potential bottleneck problems, is whether the total capacity including standby equipment represents a balanced array of productive capital goods. For, as has been pointed out elsewhere,[12] any specific scarcity within the steelmaking establishment can act to cancel the effectiveness of existing capacity because of this internally contained bottleneck.[13]

Another qualification of capacity figures is in terms of the existence of some form of rationing or allocation. A bottleneck condition which might exist were distribution patterns kept relatively fixed, might well be completely eliminated were the output of the industry channeled into certain consumer areas. In a situation where allocation (administered by the industry, for example) was used, the effective capacity of the industry could be increased many fold from the perspectives of the particular recipients of allocated output. This is one answer to difficulties in obtaining the benefits of certain plant expansion at particular times because of gestation lags. Allocation can provide selective relief to a shortage in steel which in turn might play its part in forestalling a downturn.

There is an issue associated with economic growth and change in distributive shares that should be referred to briefly. It is conceivable that a policy focused on the steel industry doing its share—even leading the way in the economy's pursuit of growth—may spill over into the distributive share arena with far-reaching consequences. This may happen on one of two fronts: first inordinate profits (particularly as a pattern setter for other industries), although a spur to growth in investment, may lead to underconsumption problems. Of course, the converse—inadequate profits—has the associated carry-over to

12. Cf. Dunn, "Report to the President," also testimony of De Chazeau, U.S. Congress, *TNEC Hearings,* 76th Congress, 2d session, part 19, November 6–10, 1939 (Washington, Government Printing Office), p. 10464.

13. In his article on steel capacity, M. Barloon stresses that it is the practice of steel firms to maintain balance in their expansion program. Cf. his "The Question of Steel Capacity," *Harvard Business Review, 27* (1949), 209.

business cycle problems, but this is removed from what is under discussion at this moment because, with low profit rates, it is unlikely that the industry would be in the vanguard of growth industries.[14] The second spill-over into distributive shares rests with the wage-price-inflation impact on the expanding economy.[15] As a part of this growth versus stability issue, it should be noted that a flat prescription of growth for the industry, instructions to "keep in line" with growth in the economy, may contain as part of the package, a pattern of increasing prices needed to keep ahead of increasing costs which will provide a savings ratio sufficient to finance this growth.[16]

Though perhaps beyond the bounds appropriate to our discussion, reference was made above to "the cultural configuration" of contemporary American society. In one sense the relevance of this topic is all too obvious. Economic policy directives cannot readily be isolated from the effects that change here may impose on a larger cultural context. But there is feedback also. A change in this roster of roles and in the associated income differentials will alter the allocation of skills and personality types, and the spread and level of aspirations in the community.

The situation is not unlike the electromotive force that develops when there is a difference in voltage potentials between two points in an electrical circuit. A reduction of the differential between two points reduces the force and, insofar as it is supposed to produce a certain torque (for example, in the case of motors) a loss of power is experienced with the reduction in the differential. In a form of economic organization in which entrepreneurial action provides the torque, the motivational force may

14. It should be noted here that the words "inordinate" and "inadequate" carry with them a notion that there is an equilibrium value; this is, of course, at the center of much labor-management controversy.

15. Cf. Otto Eckstein and Gary Fromm, "Steel and the Postwar Inflation," study paper no. 2, prepared in connection with the Study of Employment, Growth, and Price Levels for consideration by the Joint Economic Committee, Congress of the United States, November 6, 1959, 86th Congress, 1st Session (Washington, Government Printing Office, 1959).

16. Savings ratios in the aggregative sense, not necessarily in terms of retained earnings alone.

well be a function of income differentials; this, besides the issue of adequate funds from the savers of the society. This also subsumes income differentials under a broader heading which includes other motivational stimuli. Nevertheless, as suggested in the statement above, a constraint in formulating some ideal context for growth is that some precaution be taken to preserve the tone of the cultural configuration lest socio-psychological feedback deflect some of the necessary force from the system.

There is, of course, the other side of this behavioral problem. Motivational forces may give rise to so-called "undesirable" power positions and practices.

There is a question yet to be answered—and it is a bridge between the micro and macro aspects of our subject: Can we assume that recalcitrance on the part of a given firm's management to expand capacity will have impact on the aggregate performance of the industry in meeting society's need for its output? An answer lies in part with the market structure of the industry and with oligopolistic behavior patterns which will affect the transmittal of individual firm behavior to the economy as a whole. Leadership, entry, and market share factors come to the fore here.

If, in the absence of action on the part of one or several member firms in the industry to meet increased demand, there are other firms within or new arrivals from without to fill this new demand, the effect on the economy may be to provide an "adequate performance." However, if barriers to entry and leadership control, mutual interdependence recognized, implicit (or explicit) market sharing agreements, or the like are present, it is conceivable that a no-response decision within one firm may set the pattern for the industry. If so, here would be a manifestation of a most unwelcome aspect of oligopoly as far as the problem under discussion in this volume is concerned. And here also would be a logical point of departure for policy recommendations in this area. Although it may truncate the discussion, it does not seem advisable to investigate the whole range of reasons why behavior criteria have developed in antitrust litigation, or the merits of the "potential to control" issue, which has become dominant in judgments recently. It is important, however, to assess the potential of the industry to withhold adequate capacity from the economy. This is one of the sensitive elements in the power context of concentrated industry of relevance to this discussion. The following section will deal with this area of concern.

IMPLICATIONS OF MARKET STRUCTURE FOR POLICY FORMATION

For the economist, these aspects are often termed "the monopoly problem," and treated within a framework of formal analysis of the firm. He develops conclusions which set such factors as efficiency, price policy, and fulfillment of demand against environmental conditions reflected in departures from or modifications of "pure" types, i.e., competition and monopoly. Insofar as they involve normative values axiomatic to the analysis, they carry a burden which may be self-defeating. Nevertheless, this is the conventional treatment within the field, and therefore a brief exposition of this approach seems appropriate. The model which follows is modified to coincide more closely with "reality" in the steel industry; it will serve to test whether the implicit theorizing in this approach—theorizing on entrepreneurial behavior—is sufficient. Does it embrace *enough* reality? The model is illustrative of an economics-of-the-firm treatment and is a desirable point of departure for the policy material which follows it.[17]

The assumptions in Figure 10 are that a firm faces constant costs over the relevant range of production and furthermore that this cost condition (*PAC* and *PMC*) coincide with long-run average and marginal costs (*LAC* and *LMC*) which are also presumed to be linear over a continuum sufficiently long to assure that the firm could not realize reduced costs by modifications of the basic plant structure; also that additional plants established by the firm would experience cost conditions as

17. For a rigorous elaboration of the relation of "firm" analysis to investment problems, cf. Lutz and Lutz, *The Theory of Investment of the Firm.*

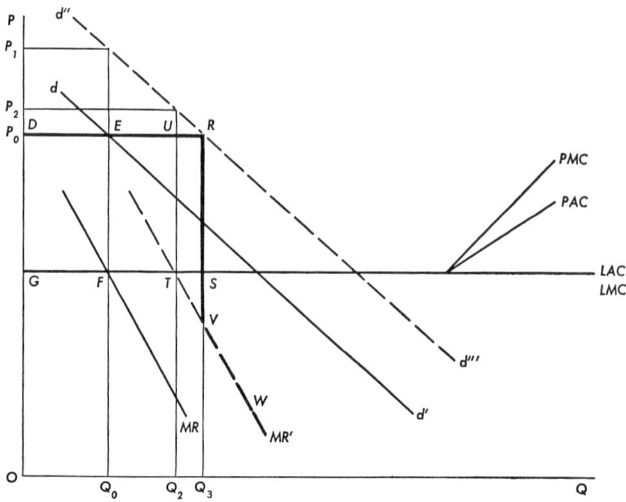

Figure 10. Output Response to Demand Change

represented in the diagram.[18] Demand conditions faced by the firm are represented by the solid average revenue line dd' and the solid line marginal to it, MR. Price P_0 and output OQ_0 represent equilibrium positions for the firm. (The issue of whether industry conditions are such as to result in a Chamberlinian equilibrium for the firm may be set aside in this example as it is not the initial condition situation that is to be emphasized but rather subsequent adjustments, once the initial equilibrium is disturbed.) Now assume an increase in demand faced by the firm as represented by broken line $d''d'''$ (with the broken line marginal to it, MR'). Possible reactions to this altered condition are: (1) the conventional new $MC = MR'$ adjustment leading to a new price P_2 and output OQ_2, which may be assumed to be unlikely because of both institutional restraints (to be discussed below) and production rigidities (at least insofar as relatively short-run adjustment is concerned); (2) a continuation of the previous output level OQ_0 and a realization of price P_1 permitted by the newly advanced demand condition; this is a more likely outcome in view of the production rigidity factor, but

18. Cf. Meyer and Kuh, "The Investment Decisions," pp. 200–02.

once again unlikely in the face of institutional restraints; (3) a maintenance of the original price P_0 and a subsequent enlargement of output as "conditions permit."[19] This alternative depends, of course, on the ability of the firm to make some adjustments as to output, either because of under-utilized (excess) or standby capacity or because we admit a time interval sufficient to overcome production rigidities. (The typical excess capacity operating rate will be discussed below.) The question, therefore, in this case is to what new level of output will the firm move? It is clear that the new demand conditions at price P_0 will allow an expansion to OQ_3, and at first glance it may appear that an output of OQ_3 will result in some production being produced under conditions where $MR' < MC$ (area STV). It may appear that, given the fixity of price P_0 the maximum output which will be attempted will be OQ_2 where $MR' = MC$, the motivation being that returns here exceed returns under the former equilibrium condition (i.e., area $DEFG < DUTG$). This, however, is *not* the case. It should be noted that insofar as the firm is operating under an imposed price ceiling (institutionally imposed), the effective demand curve will be horizontal throughout the range where the tendency might be (under less constrained oligopoly conditions) to raise price and constrict output; also that it will be downward sloping where output could be sold only at prices below the fixed price. Therefore, the effective marginal revenue curve in this situation is really $DRVW$. And the firm, with prices fixed at P_0 would produce OQ_3. Put another way, in this case—prevented from optimizing by raising price to P_2 and restricting output to OQ_2— the firm looks at expansion beyond OQ_0 as marginal gain throughout the range Q_0Q_3. Only beyond OQ_3 does it begin to experience marginal loss because only beyond this point does it face marginal costs greater than its effective marginal revenue (beyond OQ_3 MC lies above $DRVW$). Therefore, rather than being an area of loss O_2Q_3 is an area of

19. There is a fourth possible reaction, i.e., to take no action whatsoever—in effect to ignore the new demand. We assume here, however, that they will attempt to adjust output to clear the market.

marginal gain (*URST*). The significance of this is far-reaching, for insofar as the "price-ceiling condition holds," the firm will tend to expand output in the face of rising demand, up to the point where increasing costs set in, rather than, what may have been expected, to have a tendency to restrict output, or not respond to changed demand conditions, often assumed of oligopolistic industries.[20] It may be said this represents the maximizing behavior, given the price constraint, for the firm, and in a social welfare context represents a perhaps unexpected response pattern. This, of course, refers to the fact that at least as regards price P_0 (accepted by the firm, for whatever reason), demand for the product will be satisfied in a manner consistent with the market conditions reflected in the newly shifted demand curve.[21]

Throughout the discussion above, reference has been made to institutional restraints which hold the price of the product constant. This characterizes, in the extreme, price behavior in the steel industry. Relaxing this assumption to admit some upward price movement only tempers the conclusions to the extent that it approaches complete responsiveness to the new demand-price logic dictated by the changed conditions. Note that this does not suggest in any way that either the initial conditions, or any of the alternative possibilities open to the firm as a result of changed demand, depart from the Chamberlinean context of an oligopoly condition; but there is an implication that with this pricing pattern, the oligopoly firm tends to act as a pure competitor through a *part* of the demand schedule. A

20. Output is still restricted relative to what would occur under conditions of perfect competition. The shift in demand however, does elicit an output increase.

21. Meyer and Kuh add a corroborative point: "Where entry, even though difficult, is distinctly possible it behooves firms already in the industry to act on the assumption of quite high long-run demand elasticities. It is therefore to their advantage to increase output rather than permit new entrants to gather the benefits. In actual practice, the availability of expertise and awareness of the profitable possibilities will usually be concentrated in those firms already in the industry so that outsiders will be able to take advantage of the changed market situation only with a lag. If existing firms do not expand, they may find an actual downward shift in their own demand curve because other firms (either new entrants or those already in the industry) have expanded, and weakened the unexpanded firm's market position" (*The Investment Decision*, p. 201).

point to be established, of course, is the "realism" of the assumption that price rigidities (not to be confused with price identity) exist.[22]

There are forces at work in the industry which suggest that, even for reasonable periods of time, the firm may resist accommodation of the increased demand, that it might adhere to case 2 above. This would be due to the indivisibilities characteristic of the industry, as well as production rigidities.[23] The behavior here appears to represent a kind of tension build-up which at some point reaches sufficient force to break through the resistance to expansion. This build-up is not only a function of the steam (or speed) associated with advancing market demand, but also of the size of the "pressure compartment." This size is affected by (1) the entry conditions of the industry and (2) the relative fixity of cost conditions. This latter factor would alter the conclusions of the model, if costs were not linear. The trigger to increasing output might advance to a point where a firm could exploit the situation and enjoy reduced costs either by expanding capacity or by cost reduction through technological innovation or other means of increasing productivity. But given a rate of technological advance which is slow relative to secular increases in demand, this may, nevertheless, result in conditions of "satisfied" demand, insofar as pressures to keep prices stationary persist.[24]

We then say that price rigidities do not preclude expansion

22. For a discussion of studies made on statistical cost curves, cf. H. Staehle, "The Measurement of Statistical Cost Functions: An Appraisal of Some Recent Contributions," *American Economic Review, 32* (1942), 321–33, esp. references to attempts by T. O. Yntema in TNEC Hearings. For an analysis of problems in the construction of empirically derived cost curves and a specific criticism of the alleged linearity of cost curves for steel, cf. C. A. Smith, "The Cost-Output Relation for the United States Steel Corporation," *The Review of Economic Statistics, 24* (1942), 166–76.

23. A distinction should be made between the production rigidities referred to above and indivisibilities; in the former, the reluctances to expand capacity stem from motivational factors as well as the complications and difficulties inherent in planning for and achieving additions to output. In the case of indivisibilities, reference is specifically to the so-called "lumpiness" of capital facilities in an industry such as iron and steel.

24. Note that in this latter aspect the pressures which lead to *holding* the price line, insofar as they emanate from without (e.g., the Kefauver Committee), apparently give rise to a pattern of satisfying demand.

in the face of rising demand. We have not said, however, that investment would not occur also under competitive pricing. Nor have we argued the issue of whether competitive pricing (and associated freedom of entry) would solve other problems, e.g., would remove a source of inflationary pressure,[25] or reduce an "inordinate" concentration of market power.

Our model rests implicitly on limited entry. It was noted that firms were likely to move themselves to satisfy demand to bulwark the entry barriers. Therefore, if market structure and pricing policy are judged solely on the basis of whether the circumstances present logical reasons for expansion of capacity, conditions of price rigidity and limited entry may qualify.

With regard to these two dimensions, at least, we may conclude that conditions in the industry may very well be left as they are. Where, then, do things possibly break down? What we have said springs from an essentially static model. It completely neglects the question of timing. And, as we have emphasized in earlier chapters, it is the inherently dynamic character of investment *demand* and the cyclical problems in the over-all economy that are of concern here. We are still left with the acknowledged indivisibilities of physical capital in the steel industry which prevent gradual and smooth adjustment to changing demand conditions. And we are still faced with the question: Can capital formation in steel be regularized? Can we rely on secular excesses or hope for more accurate planning to minimize contingencies?

Part of the answer can be drawn from materials in Chapter 5. We must make several additional observations on managerial motivation.

MANAGERIAL MOTIVATION

In addition to internally generated ideas about their responsibility for producing goods and maintaining profitable opera-

25. There are some who maintain that steel may at times be underpriced. It is also suggested that under highly competitive conditions the industry might well be "unprofitable" over the cycle, confronting the economy with the need for subsidies, etc.

tions, and in addition to a set of goals partially arising in the pursuance of the so-called operational objectives of their work, there is a set of identification factors which may be characterized as the *commitment* which the individuals of the organization feel. This is a commitment to the goals and to the ideology (arising both from within the organization and "imposed" from without) of the enterprise of which they are a part. This is an aspect of their psychological state which has relevance for our discussion because it helps to explain modifications of behavior, departures from the single-purpose profit-maximizing directive.[26]

The concept of commitment has both positive and negative attributes. On the one hand, commitment to the value system of the organization inhibits the range of admissable alternatives open to the entrepreneur, in that in the absence of this commitment, he might be willing and in a position to consider a range of alternatives for action much broader than that which is now open to him with this constraint. The idea of commitment suggests that the decision-maker is tied to a set of policies or precedents which confine his action.[27] Very often this means that

26. "A commitment in social action is an enforced line of action; it refers to decisions dictated by the force of circumstance with the result that the free or scientific adjustment of means and ends is effectively limited. The commitment may be to goals, as where the existence of an organization in relation to a client public depends on the fulfillment of certain objectives; or, less obviously, to means, derived from the recalcitrant nature of the tools at hand." P. Selznik, "A Theory of Organizational Commitments," in *Reader in Bureaucracy,* ed. R. K. Merton, A. P. Gray, B. Hickey, H. C. Selvin (Glencoe, Illinois, The Free Press, 1952), p. 198; cf. also pp. 199–201.

27. In fact, it has been suggested that loyalty to the organization provides the rationale for dissociation with larger social responsibility. Boulding states: "I know of no instance in history . . . where a group or an organization has voluntarily laid down its life in the interest of humanity. In the defense of their nation, their church, their union, their business, their family, men have been known to lie, cheat, steal, and even murder with a single-mindedness of intent that an individual as such rarely achieves. A divergence of private and social interest is all the more serious, therefore, when the private interest concerns a group within society, even where that group is as large as a nation. It is a divergence of interest against which moral exhortation is singularly ineffective, because the fact that the individual is serving *some* group which is greater than himself blinds him to the fact that his group is only a part of the whole" (*The Organizational Revolution: A Study in the Ethics of Economic Organization* [New York, Harper, 1953], p. 220).

once embarked on a particular line of action or an enunciated set of values or a particular characterization of role, the individual is bound to proceed in the direction of this suggested line further than might have been envisaged at the initial stage of forming policy or objective. As Selznik has stated:

> The systematized commitments of an organization define its character. Day-to-day decisions, relevant to the actual problems met in the translation of policy into action, create precedents, alliances, effective symbols, and personal loyalty which transform the organization from a profane, manipulable instrument into something having a sacred status and thus resistant to treatment simply as a means to some external goal.[28]

The negative aspect of the commitment concept, then, rests with a circumstance which limits free range in choice of goals and means of realizing goals, a circumstance which enforces "uniformities of behavior."

On the other hand, the question, "to what are we committed?" carries with it a suggestion of dedication which may in itself be a source of motivational energy and which may result in drives to pursue a certain line of action, ending in a measure of success which might not have been attained in the absence of the commitment that accompanies action. In certain situations, it is this latter aspect which manifests itself as "socially responsible" behavior of entrepreneurs. This psychological state is also responsible for organizational loyalties which lead, in turn, to responsible planning and action for the long run, planning which in some social welfare sense has implications for a desire to stay with and foster forces of expansion in the economy, rather than a cut-and-run outlook leading to instability and disruption.

An accompanying factor is the desire for and concentration on the continuity of the enterprise. The managerial group may well entertain, as a result of commitment to the enterprise or

28. Selznik, "Theory of Organizational Commitments," p. 201.

because of other ideological ties, a desire to insure the persistence and well-being of the organizational entity. This may well lead to a roster of goals wherein continuity of the enterprise holds a higher priority than the desire to keep the stockholders satisfied with attractive rates of return or the public at large with socially acceptable performance—i.e., measures designed to keep these groups contented. The priority list may place continuity of the enterprise, as a goal, ahead of personal financial reward to the executives themselves (witness recent modifications of bonus and salary payments to executives in the face of other considerations confronting the firm). It is not necessary here to rank these goals, but merely to suggest that in given situations, in given organizations these goal rankings will vary and that it cannot be blandly assumed that profits, as such, serve for some purposes at least as more than a kind of generalized expression of the complex motivational pattern that exists.[29]

The most typical alternative goal to profit maximizing enunciated as such is the claim that managerial enterprise desires the ascension of power and the consolidation of power both in terms of the internal organization of the firm and in terms of a position of dominance in the economy as a whole. Consolidation of power is often explained in the simplistic terms of aggrandizement of personal ego, but a fuller analysis of this goal relates the physical or technological conditions which lead to the development of large-scale operations and the possibility that concentration of control may serve as a hedge against the great uncertainties associated with economic activity. With a trend first toward concentration of power and later to a pervading characteristic of oligopoly in the economy, there has developed a social or public reaction to this power and a desire

29. Selznik has suggested an even broader concept of self-protection for the firm referring to a need to safeguard "the security of the organization as a whole in relation to social forces in its environment." Cf. P. Selznik, "Coöptation: A Mechanism for Organization Stability," in *Reader in Bureaucracy*, pp. 135 ff. A desire for security, it has often been suggested, has led management to play it safe and in the face of uncertainty not to pursue profit-making opportunities with the degree of vigor that might occur in a more "knowable" environment.

for public control. A fear of abuses of power, a fear on the part of those not in control in a political environment where their apprehensions may find expression through legislative action, and an accompanying image of the idealized operation of a capitalist economy as one where unhampered operation of the market mechanism performs functions of allocation and the maintenance of equity in the society—have led to an interplay of forces affecting both the goal structure and ideological format of enterprise. And the perhaps ironic accompanying development has been a felt need on the part of the emerging corporate bureaucracy to seek protection, in selected areas, from the vagaries of the uncertain economic environment in the very agency which is given by others the task of controlling the behavior of the "offending" organizational bureaucracy.[30] As various groups in the society (labor and consumers as well as business) look to government as the source of control and arbitration, the power of, and the need for, governmental bureaucracy develops: a juxtaposition of power blocks in the society ensues. Of course, business leaders want a certain amount of "interference," but not *too much*. "It is clear that the various business groups must continuously seek to draw and redraw [the line between the provision of safeguards and some point beyond this], notwithstanding the basic ambivalence of their whole position in its economic and psychological aspects."[31]

With all this, however, and although we may assert that managers of large enterprises do have a strong sense of social responsibility which leads them to pursue courses not consistent with short-run profit maximizing, we have not pushed profits from the stage. The expansion in steel capacity of recent times

30. For nineteenth-century experience with this apparent paradox, cf. my "The Role of the State in American Economic Development, 1820–1890," in *The State and Economic Growth*, ed. H. G. J. Aitken (New York, Social Science Research Council, 1959).

31. "Thus, what appear before as the more or less unanimous voice of 'business' in its opposition to governmental interference may then turn out to be a struggle among the different groups of business and industry—with, in, and against the government, with as well as against other interest groups—over the incidence of political power and economic policies" (R. Bendix, "Bureaucracy and the Problem of Power," in *Reader in Bureaucracy*, p. 126).

in response to government and public prodding is an example of this point.[32] P. L. Bernstein says, in this connection:

> On the surface, these policies seem to be more in line with operating 'in the public interest' than with the traditional concept of profit maximization, for there is no doubt that they cost money in the short run. *The important point* is that, over the very long run, profits are being maximized in the sense that a less responsible attitude on the part of management could lead to very severe repercussions in the form of public hostility, anti-trust actions, and nationalization.[33]

What emerges from study of both pricing and investment behavior is a pattern of maximization within *perceived* limits. Commitment to certain non-pecuniary objectives may temper the course chosen and "maximization" may proceed only part-way because of decision-makers' awareness of potential control, interference or censure.[34] This may mean that the course chosen and the end-point sought do not result in absolute maximum money returns; it may indeed also mean that action taken is "socially responsible." But it is similarly true that neither of these things may happen.

Much hinges on perception and commitment. If the industry

32. Eisner quotes a steel executive in the course of an interview on expectations: "We came to realize about 1949 that we were on a new plateau and that we had to expand steel capacity. It took four years to accept this expectation." See *Determinants of Capital Expenditures: An Interview Study* (Urbana, University of Illinois, 1956), p. 55.

33. P. L. Bernstein, "Profit-Theory—Where Do We Go From Here?" *The Quarterly Journal of Economics, 67* (1953), 415.

34. A number of formulations alternative to simple maximization have appeared in our discussion, e.g., those of Papandreou, Simon, and Knauth. Cf. also W. J. Baumol: "My experience, and apparently that of some others who have worked with business firms, is that profits do not constitute the prime objective of the large modern business enterprise. . . . I shall maintain that the size of the firm's operations shares with profits the role of prime objective. . . . I am prepared to generalize from these observations and assert that the typical oligopolist's objectives can usefully be characterized, approximately, as sales maximization subject to a minimum profit constraint" (*Business Behavior, Value and Growth,* New York, Macmillan, 1959, pp. 31–32, 49).

misjudges public temper, through lack of insight or through commitment so at variance with social goals, it may not be able to pilot a course that assures sufficient accommodation to outside forces. It may flounder on barriers which tear at its sovereignty or self-proclaimed just rewards.

From the public's viewpoint, we may welcome industry demonstrations of socially responsible behavior and at the same time be apprehensive about possible social costs; that is, possible misallocation of resources due to diversions from pure profit maximization.[35] But more than this, when we do have "socially responsible" decision-making and only this, we have something given to us out of largesse from a private sector of the society. We have no guarantees, nothing to fall back on should the leadership shift *its* goals, perceptions or evaluation of what is in its best interest or the public's best interest. In sum, we are saying that though "enlightened oligopoly" may come through, it is not certain. Can we increase that certainty?

SOCIAL CONTROL MEASURES

Before moving to a list of measures which might affect industry decisions, the question of the discrepancy between estimates of needed capacity arising from within the industry and those from government and other "external" observers should be recalled. The issues and factors involved have been treated at length in Chapter 4. Building an analytical frame for work in this area, developing methods and procedures, and actually acquiring needed quantitative materials is underway.[36] We need only re-emphasize here that as methods of demand analysis and projection improve—both *within the industry* and among outsiders devoted to this work—one dimension of the problem will become of less significance. As the discrepancy in quantitative

35. Inhibitions within management which lead to reduced short-run profits may well rob the system of adequate capacity. Cf. Ralph C. Jones, "The Effects of Inflation on Capital and Profit: The Record of Nine Steel Companies," *The Journal of Accountancy, 87* (1949), 12.

36. In the work of Koopmans, Leontief, Klein, Orcutt, Rand Corporation, et al.

estimates narrows, other areas of divergence, such as appropriate goals and questions of equity, may be faced with clearer perspective.

We must begin by acknowledging two aspects of the control picture which reflect the actual state of affairs. On the one hand, we recognize that the steel industry is something less than a completely private industry. It faces today, as in the recent past, extremely close scrutiny from members of Congress, the executive, the courts, and from the public at large concerning many of its activities. This is, of course, what lies behind the much discussed "social awareness" within the industry. And, as we have indicated, there have been menacing thrusts in the direction of the industry's structure, some in the form of statements favoring the dissolution of large units in the industry. In one case there has been a threat of government entry into actual production.

We must set against this the second aspect, namely, the fact that there are limits to controls and the intrusion of public regulation or influence. We do have the corporate form, and organizational enterprise is a dominant, controlling organism on which we depend for the provision of needed commodities. The business corporation has its ramifications not only in its immediate dedication to production but in the somewhat more remote, but nonetheless compelling, aspects of our way of life. Proposals for drastic changes in the character of workings of the corporate form carry with them inevitable implications of tearing at the larger social fabric.

A roster of alternative measures might be suggested (and, indeed, in various forms all have been suggested)—a series of measures which range from extreme moves to enforce competition to proposals which take for granted the public utility aspect of the industry's situation and maintain that, in fact, the industry should be handled as a public utility. Some, therefore, might be found in a column headed "More Competition," others in a column headed "More Control." At the outer fringes in one direction is complete dissolution, where standards are put forth as to what is a maximum size firm (very often using a fully integrated works on one geographical site as a standard). In the other, there are discussions of nationalization and formalized government competition with the industry in a manner similar to TVA in the electric power industry. Within the boundaries set by these extremes lie a number of institutional modifications or rearrangements. In some cases, they involve the employment of devices already used in other connections to cope with the problem of industry behavior. Some of these may provide an answer to the questions asked in this volume regarding problems of expansion of capacity. A brief catalogue of these, before proceeding to some explicit proposals will be useful.

Accelerated amortization has been used both during World War II and during the Korean War to induce industry in general —and consequently the steel industry as well—to expand capacity. Though the general aims and relative effectiveness of this device are beyond our discussion,[37] it might be well to cite the economic effects claimed for it:

> They (1) encourage the expansion of facilities on a selective basis in accord with broad national objectives; (2) increase the tax revenue base, through expanded revenues from increased sales, while decreasing the tax revenue from each sale; (3) expand the privately-owned industrial plant of the country relative to the government-owned portion; (4) increase the rate of turnover of term capital and, in that sense, increase the amount of available investment funds; (5) increase the dependence of private industry upon the government for the assumption of risks; (6) increase the rate of obsolescence of existing facilities by speeding up the rate of construction of improved plants and equipment; (7) discourage the flow of investment into term projects, which are not sufficiently essential to merit receipt of such special tax arrangements; (8) increase the pressure on

37. Cf. J. P. Miller, "The Pricing Effects of Accelerated Amortization," *Review of Economics and Statistics, 34* (1952), 10–17.

> price control authorities and procurement authorities to recognize the "higher costs" of overhead during an emergency; and (9) increase sharply the price pressure at the basic materials level, at least through the period of accelerated use of rather than production from the expanding facilities.[38]

In his remarks on this, E. C. Welsh claimed that the steel industry was very definite in asserting that it would not expand capacity without being able to take advantage of accelerated amortization. The government was able to use this in a form that permitted selective choice of what aspect of the productive process would be expanded in relation to other stages. In fact, it concentrated on the development of ore facilities and coke ovens rather than on finishing facilities because the latter were more abundant. The government was able to accomplish this by allowing higher percentage rates of amortization for the coke and ore stages.

In general terms, those who support accelerated amortization as a tax policy measure do so on the grounds that it involves a minimum of intervention in the private sector and that it has a commendable quality of administrative flexibility. And yet when we find it applied on a selective, allocative basis as in the Korean War, it surely takes some freedom from the private decision-maker. It also carries with it the possibility that firms are more likely to take advantage of fast write-offs during periods of high demand, thereby intensifying their liquidity position at a time when there are reasons to hold this in check. Perhaps most damaging is the concern that a policy of accelerated amoritization tends to favor firms already in existence, firms which will avail themselves of an opportunity to round out facilities. It does this rather than afford to potential new entrants in an industry the means to finance such a venture.

There are other tax proposals which apply more directly. Though they may be "discriminatory," they nonetheless afford

38. E. C. Welsh, "Government Aid to Business Expansion," *The American Economic Review, 42* (1952), 423–27.

the society an opportunity to encourage investment in particular areas. Among them is the granting of tax credits. And this approach does not necessarily mean singling out particular industries for favorable tax treatment; it may mean across the board tax allowances for investment which go beyond replacement and genuinely involve expansion of new facilities. But here, as with so many other proposals, we face the implicit danger of moving further and further away from the price mechanism and its functions regarding allocation in the economy. A recurrent theme intrudes—indeed, an idea expressed in the introductory materials of this volume: do we recognize that there are social costs involved in particular tax programs designed to direct investment activity and then do we move to the conclusion that, set against what appear to be larger social costs, we are justified in considering such devices?[39]

What might be deemed a more affirmative approach to aiding industry, in the face of a desire to expand capacity but an inability to do so because of inadequate financial resources or access to financial markets, is found in proposals which suggest some program of government lending. Once again, however, the presumption would be that loans would be granted on some basis of externally determined "desirable" investment activity. It might be possible for the government to influence decisions as to levels of capacity and the product mix chosen by the industry through an office and a set of procedures similar to those used by the Business and Defense Services Administration. If done in this fashion, priority directives would be issued which might well have effect not only on the position of individual firms, but undoubtedly would affect pricing policies as well. It seems likely that government loans would have their greatest impact in making funds available to firms who do not have internal resources to finance expansion and therefore, they would undoubtedly affect entry conditions in industry. Such loan activity on the part of government could also be used to

39. Cf. W. J. Baumol, "Proposals for Increasing the Growth of National Output," in K. Knorr and W. J. Baumol, eds., *What Price Economic Growth?* (Englewood Cliffs, New Jersey, Prentice Hall, 1961), especially pp. 30–34.

enhance a program of regularization of investment activity and to accommodate some over-all counter-cyclical economic policy. The real problems in this approach seem to lie largely in the bureaucratization of government loan procedures and decisions which tend to lag changing economic conditions, which tend to be generally unresponsive to these changes. Complications arise in the inflationary side of the cycle when government actively supplements the expansion of loanable funds occurring in the private money market and may come into conflict with other government counter-cyclical policies. As with accelerated amortization and tax credit arrangements, government loans have become an established component of government programs to direct and encourage investment activity.

The fact that the government is a final buyer of industrial output of enormous scale is a feature which has loomed large since World War II and shows no sign of diminishing. Government leverage in directing and controlling product mix as well as the amounts and directions of new capital formation is of tremendous significance. This may well prove to be one of the most potent devices available to the society to affect investment behavior. Indeed the position of the government as a discriminating monopsonist may prove to be a wedge, in the direction of controlling the private sector of the economy, of such proportions that we face in this one development serious philosophical questions. The de facto presence of government's ability to influence crucial decisions has occurred without the society ever examining, in a fundamental sense, the desirability of, or its willingness to have, this quite radical modification of the economic environment.

If industry action is so intimately bound up with the public interest and if, as some have asserted, the industry does in fact act as if it were a public utility, then we might ask should not the industry be treated as public utilities are treated? Should not the industry's behavior be regulated by an independent commission? It may be argued that though the problems associated with utility regulation are manifold, so too are the social costs associated with the investment problem discussed in this volume. Therefore, as one author has put it: "Might not the 'rigidity of a politically responsible regulatory body' be preferable to that of what Berle might term a non-statist civil service, comprised of managers answerable, at best, only remotely to their stockholders?"[40] But what of the effectiveness of this form of social control? A comprehensive study of independent regulatory commissions, after a detailed examination of the relative success of various efforts in this field, comes to a set of negative judgments about the effectiveness of regulation by commission, in terms of administrative inefficiency, lack of effectiveness in accomplishing its purpose, and its general failure to make a creative contribution to the economic environment. The study concludes with the following remarks:

> The persistence of myths about the qualities of independent commissions endangers the achievement of effective regulation of business in the public interest. Commissions have proved to be more susceptible to private pressures, to manipulation for private purposes, and to administrative and public apathy than other types of governmental organizations. They have lacked an affirmative concept of the public interest; they have failed to meet the test of political responsibility in a democratic society; and they tend to define the interests of the regulated groups as the public interest. The effectiveness of regulation of business by commission hangs by a thin thread.[41]

Perhaps, then, government competition with the industry of the TVA variety would have advantages over regulation by commission. It surely would establish a standard of value and

40. I. M. Stelzer, "Review of *Market Power: Size and Shape Under the Sherman Act* by G. E. Hale and R. D. Hale," *The American Economic Review, 49* (1959), 1110.

41. M. H. Bernstein, *Regulating Business by Independent Commission* (Princeton, Princeton University Press, 1955), p. 296. See also W. Adams, "A Critical Evaluation of Public Regulation by Independent Commissions: The Role of Competition in the Regulated Industries," *American Economic Review, 48* (1958), 528–29, 542–43.

provide information which then might be used to make judgments about the performance of the private sector. It seems likely, with government competition, that private managers would be more responsive to opportunities to improve efficiency, take advantage of new developments, expand into new markets, etc.—more so than they would were many of their traditional areas for decision circumscribed by government control. On the other hand, industry has been very restive in the face of TVA-type competition. It is particularly sensitive to the possibility that government enterprise might receive subsidies which do not appear on the surface and might, in fact, be able to operate from a different risk base because of a favored position as part of the federal economic establishment.

Yet, with the potentially preferable aspects of the public competitive corporation as one possible means of affecting industry behavior, this does represent as radical a departure from existing institutional arrangements as one could think of short of actual nationalization of the industry. Indeed, if this measure were to be imposed as an answer to some of the problems we have discussed, we might well find a resulting modification of the basic economic format that goes well beyond what the situation in fact might require.

Is there some middle ground which, though calling for some changes in the set of institutional arrangements, still permits our living in essentially the same economic environment that we now have? Are there solutions to this problem which allow us to turn away from the radical surgery of dissolution and from equally radical moves in the direction of government ownership of facilities or government competition with the industry in a direct manner? Can we induce the holders of considerable market power to act with restraint, wisdom, progressiveness, and a sense of what the larger social needs are? This takes us to what is referred to as control by admonition.

What has emerged from our examination of decision-making within the industry, from our general characterization of the entrepreneurial group within the industry, and from our reflections on the effect of the sense of commitment and perception that controls in the decision-making process within the industry —suggests that control by admonition may have its merits and may have much to recommend it. An examination of recent attempts of the government to elicit behavior in the labor-management field, however, leaves the impression that more than admonition, conducted on a very personal and very intense level and brought forth on a conspicuously *ad hoc* basis, is necessary. The missing ingredients in regulation by admonition stem in large part from its potentially capricious nature. And it is in the light of this skeptical reaction to the effectiveness of such measures as they are presently pursued that we will now turn to a set of specific proposals.

We have rejected as unrealistic the possibility of recapturing a world of atomistic competition. Even if the government were to embark upon a program of changing the number of industry units according to a measure which set the maximum size at one integrated works on one geographical site, the number of decision-making centers would be enlarged, at best, from perhaps eleven to something still under a hundred. Yet we have established that even within an oligopolistic structure it is not unlikely that private decision-makers may, in the face of perceived increasing demand, be willing to expand capacity. Nevertheless, as we saw in an earlier part of this chapter, an apprehension remains regarding the timing of such an expansion. Will it be available when the society needs it, if it is to be provided at all?

In rejecting as a remedial policy the strict dissolution of large concentrations in the industry, however, we may still take a stand in favor of keeping entry conditions as open as possible with an eye to the future. This can be accomplished in part at least on two fronts: the society can take a firm and unwavering stand with regard to moves from this point forward toward further consolidation of power, i.e., it can oppose any further mergers within the industry. Secondly, it can make financing readily available to any from outside of the industry who, as a result of the attractiveness of the financing arrangements and the profit-making possibilities within the industry, choose to enter.

It is believed also that tax policy can supplement the antitrust side of the picture, resulting in conditions of more dynamic responsiveness to the economy's needs for investment in the industry. The problem posed in this study could, in part at least, be met, through the combination of a vigorous pursuit of antitrust policy and a tax credit and government loan program, readily available to all comers willing to embark on investment consistent with requirements as seen by perhaps the Council of Economic Advisors or other competent observers.

In addition, still within the realm of antitrust policy, the price level of the industry should be encouraged to advance at times when, in fact, steel is in short supply. We need a reversal of the pattern of the immediate postwar period when, in order to preserve its oligopolistic province, the industry artificially suppressed prices as a means of effectively erecting barriers to entry. A qualification to this suggestion would be to view and perhaps modify such encouragement when it would be in conflict with government counter-cyclical policy and solutions to international balance of payments problems.

As with dissolution in the discussion above, we rejected the suggestions from some quarters for public utility regulation or more extreme moves in the direction of a public corporation or nationalization. We rejected these on grounds that they, too, were neither economically, politically, nor administratively feasible nor desirable. But on this side, too, we may borrow some measures which normally would be part of a regulatory policy package. Here we refer specifically to the informational component of such programs. Without suggesting a set of production targets which would be binding on the industry or which would dictate to the industry what must be produced and at what rate, we can, nevertheless, favor the establishment of a review body devoted to such matters. This might take the form of a government agency which would work with the projections of gross national product originating in the Council of Economic Advisors, breaking these down to provide estimates of what aggregate national capacity and product mix would be required of the industry. The existence of this information, publicly available and publicly discussed, does not necessarily assume that it points the way to judgments superior to private decisions based on the internal information and the expertise that exists within the industry. It does, nevertheless, put into the public record a set of targets against which the industry as well as the society may judge industry performance.

The British have instituted similar procedures; we can find, in a description of the powers and duties of their Iron and Steel Board, some indication of what this type of agency may have as its objectives:

> The financial responsibility for development lies with individual producers who in the end must plan and carry out the projects. Development is a continuous process both as regards the modernisation and expansion of capacity. It is the subject of continuing study on the part of individual producers, their trade associations and the Board. It is the Board's function in consultation with all concerned to form a judgment about the level of demand; to consider whether the scale and pace of development is adequate in relation thereto and whether individual projects are sound in themselves; to ensure that the pattern of development is likely to conform with longer term conditions; and if necessary to persuade producers to extend or modify their plans accordingly.[42]

In the suggested public review procedure, we reverse the burden-of-proof balance that currently seems to favor industry. Whereas at present it is the government or the society in general that must come into the public forum and build its case, make its appeal, and prove its assertions that the industry is not performing adequately, the establishment of a continual review from a duly constituted agency which sets forth relevant information and imposes on the industry an obligation to explain "why it is falling short," can change important aspects of the

42. Iron and Steel Board, *Development in the Iron and Steel Industry, Special Report, 1961* (London, Her Majesty's Stationery Office, 1961), p. 1.

investment planning process.[43] More important, once established and routinized, it may well formalize the inclusion of an element receiving lip service or erratically sincere attention by the industry, namely, the meeting of larger social obligations. As things stand at present, we have no clear definition as to how felt obligations shall manifest themselves and where performance shall be put to the test. Under this new format, the hope would be that the decision-making procedure, within the private sector, will inevitably build the public hearing factor into its operations.[44] The boards of directors, in considering their plans for the future, would now not only ask, "Is it profitable, does it sustain the continuity of the firm, does it affect our share of the market?" but also, "Are we going to be questioned about the adequacy of our moves, the correctness of our moves?" As with so many questions, of course, we recognize that this proposal tampers with the completely unfettered private decision-making operation. But set against this is the benefit that social responsibility becomes defined, pinned down and delimited.[45]

The third avenue of social control, referred to as admonition, was also treated in the foregoing as something short of desira-

43. If analogy is useful, the impact would have more in common with SEC than with ICC procedures, particularly as regards disclosure, etc.

44. Note that this proposal differs from a special fact-finding commission, in that it has continuing functions. Once created it would be a factor to reckon with in planning for the foreseeable future. Its range of interests and approach, however, might be similar to that suggested in earlier proposals on fact finding. E.g., "It is recommended, therefore, that a Federal impartial and non-political fact-finding commission be established, staffed with men qualified by economic, statistical, and accounting training and versed in the complexities of the industry; that it be vested with authority to prescribe and collect the type of information required; that it be authorized to make public the information collected in any form that will not explicitly identify a given establishment; that it be directed to hold public hearings and summon thereto representatives of all affected groups; and finally, that it be charged with the responsibility of preparing, on the basis of its examination of the problem, a report to Congress with detailed and summary recommendations concerning an effective public policy toward the iron and steel industry" (Daugherty, de Chazeau, and Stratton, *The Economics of the Iron and Steel Industry*, pp. 1149–50).

45. Some line would have to be drawn as to which industry or firms come within the purview of the review agency. Among various alternatives are choosing those firms above a certain asset size for review, or choosing those firms which satisfy an amount of final demand in excess of x per cent in their industry.

ble.[46] However, as a follow-up to a program embodying suggestions in the preceding paragraphs, we may readmit this factor. We find that admonition comes as the first stage in an effective procedure and, insofar as it complements other steps taken, it is preferable to other policing measures open to the government. Admonition within this framework loses its capricious quality. The obligations, and the yardstick against which these obligations are to be measured, are established. The respective firms know they are to be judged in very strict accordance with the antitrust laws as they exist or as they may be modified. The corporation has a clear mandate as to how its social responsibility will be measured; there will be a public forum where an agency—committed to protecting the "public interest"—presents its case, and the industry, then, will come in with its answer.

Even with this apparatus, however, the industry may still say that it is its decision and its best judgment that government estimates are wrong, government judgment is wrong.[47] At this moment, the government can decide whether to unlock its bag of preferential tax treatment, preferential purchasing, and other rewards and penalties open to it. The distinction between what is being proposed here and what exists under present conditions is that in spite of its public pronouncements with regard to social responsibility, the industry just does not know against what standard it is to be measured, and it is just not sure what moves will be invoked in response to action taken. And they may, as things stand now, have good reason to claim that sudden and apparently erratic recriminations on the part of government carry with them elements of discriminatory and unfair treatment.

46. Cf. B. W. Lewis, "Economics by Admonition," *The American Economic Review, 49* (1959), 384–98.

47. "The administrator must learn that facts are not 'facts' to the resistant person; the learning of uncongenial material is slower, the forgetting faster; even when evidence is assimilated and becomes knowledge, incompatabile attitudes persist. These precepts are known when the government deals with ethnically differentiated groups; they are familiar to the students of labor-management relations; they may be assimilated now to the practices and lore of economic regulation." R. E. Lane, *The Regulation of Businessmen: Social Conditions of Government Economic Control* (New Haven, Yale, 1954), p. 119.

We have discussed a number of policy changes which might be instituted to affect the long-run steel capacity picture. One final measure is related to the Employment Act of 1946. If, as a society, we continue to hold to the full employment commitment expressed in that legislation, we have at hand a "staging area" in which to frame the foregoing policy recommendations. It may be true, in terms of materials presented in Chapter 3, that the question of steel capacity comes within the Act without any modification of its present wording. But there are reasons for including explicitly the fact that we are committed to economic growth and the associated investment goals in addition to being committed to full employment. If this were done, we would add considerable tone and importance to the administrative rearrangement suggested in our policy recommendations.[48] Such a public declaration of national economic philosophy could play a significant role in establishing the ground rules for the other institutional arrangements, recommended to assure adequate levels of investment in the economy. To be sure, the mere enunciation of this additional dimension in the wording of the Employment Act would not be enough. We have indicated that it must be accompanied by monetary and fiscal measures and by antitrust policy which conform to our objectives, and supplemented by the public hearing board and informational agency which would accumulate and disperse information concerning the capital needs of the society.[49] However, as with questions of interest rate policy or public expenditure policy, the Joint Committee on the Economic Report would review the findings of the proposed hearing board, and would be concerned with investment decisions in strategically placed industry as well as with other decisions which affect the successful attainment of stated goals.[50] To re-emphasize a point

48. This suggestion, of course, does not preclude adding reference to the desirability of general price stability as well.

49. Such changes in the Employment Act could provide the vehicle for establishing the review body mentioned above.

50. The hearings of the Joint Committee on the Economic Report would appropriately include coverage of the following questions: (1) the reconciliation of differing views as to demand conditions, (2) the implications of investment plans within the industry as they might affect over-all programs for economic stability and growth, (3) judgments concerning the "reasonable-

made earlier, it would be the regularization of public hearings on these matters which would fill out the set of recommendations we have discussed. This formalization of procedures, stemming from an already accepted format, the Employment Act, would have the additional benefits of precedent to aid it in its effective implementation.

To summarize our discussion of social control, we have rejected, for various reasons, the alternatives of nationalizing the industry, placing the government in the steel business in competition with the private sector, public utility regulation, and those types of admonition and other ad hoc procedures which are pursued with vaguely defined precepts, capriciously applied. We have taken a position in favor of vigorous antitrust policy, which has as its goal the promotion of more permissive entry conditions and the prevention of further consolidation of power in the industry. But, recognizing that a procedure of dissolution proceedings against the larger firms seems to be a politically untenable recommendation, we have turned to alternative measures which include: the establishment of a board for the dissemination of information (including more accurate demand analysis), the marshalling of fiscal measures to aid in providing adequate financial resources to those firms acting in accord with national economic policy, and the introduction, through amendments to the Full Employment Act, of regularized hearings drawing on the activities of the review board and other sources of information. The intent has been to establish procedures leading to enlightened, equitable, and regularized government response which will, in turn, affect decisions in the private sector of the economy. We have aimed at a general revision of the psychological outlook, the sense of commitment and perception within the decision-making group in the industry. We would hope that this change in outlook would include (1) an optimistic attitude on the prospects for national economic growth and stability, (2) a vigorous, affirmative commit-

---

ness" of the rates of return to industry on their investment plans, and (4) an examination of the market structure of the industry which affects investment behavior.

ment to the contribution that the industry must make in the pursuance of national economic goals, (3) a recognition of the sensitive linkage between managerial actions and subsequent public reactions, and (4) a real consciousness of the desirability of fostering and enlarging the competitive elements within the industry.

SOCIAL COSTS AND THE FUTURE

In this study, more attention has been paid to the question of potential bottleneck than to excess capacity. This emphasis may seem misplaced, in the face of what the historical record has shown to be "typical operating rates" in the industry, and also in the face of the possible new era of competition from substitute materials and from changes in international markets where the American industry may be hurt. But it is precisely because of the emergence of new conditions within the industry and because of a hoped-for increase in the pace of over-all economic development that this emphasis seems valid. The bottleneck problem may be expected to grow in relative importance in this new anticipated setting.

If we hazard a guess as to what the industry's situation will be some twenty years from now, we may visualize relatively few large domestic producers competing in an unprotected international market. Demand for the industry's output will be focused on a much higher proportion of finely designated, specialized uses. Rapid technological change demanding high rates of obsolescence and, perhaps at the same time, involving lower relative rates of capital consumption may well be part of the picture. It seems unlikely in the face of these speculations that the desirability of industry responsiveness to changing demand conditions, technological opportunities, and macro-economic requirements which we have emphasized, will be lessened.

The excess capacity situation, reflected in industry figures, coincides with the general demand pattern faced by the industry in the past and it is likely to continue on an intermittent basis. Because of its position far back on the production spectrum, the industry will continue to experience something other than smoothly changing demand conditions. The movement on the part of some firms in the economy, with idle capacity at the beginning of an upswing and investment only at various later periods, will continue to result in the transmission of uneven reaction patterns to steel markets. And this will continue to produce conditions in which capacity will be in excess in some periods. Our point, however, is that in order to assure adequate capacity at all times, we must be prepared to sustain these excesses at certain times.

If, in this sense, excess capacity is viewed as standby capacity, it follows that financing of this capacity might be looked upon as something other than a wastage cost borne by either the consumer or the industry. From this we can say that tax policy which favors the provision of such standby (or excess) capacity, or price behavior which also provides such capacity, may be viewed quite differently—if in the one case this capacity were considered to yield a social return as opposed to another, wherein it was looked on as a manifestation of inefficiency or uneconomic planning.

When viewed as a social cost that should be met, we do have a different perspective on this situation and may be more willing to consider alternative ways of bearing this cost. Whatever policy measures are pursued, however, it is hoped that they will be formulated with the recognition that the timing of investment activity in the steel industry is a matter of far-reaching consequence to the society. We know that we will have excess capacity at times. Will a program of building ahead because of the industry's potential for choking growth in the economy make these periods of excess capacity take on the attributes of a perhaps more rational, acceptable, and even necessary condition?

It is hoped that this study provides material for more complete characterization of the interaction between the individual decision-maker and the more abstract structure of the working economy, so that individuals within industry and individuals in government (who are charged with the responsibility of eliciting responses from the industry) will see more clearly where a meeting ground may lie. In terms of normative judgment, of course, it is hoped that the measures, the specific policy modifications suggested, will bring about the closest meeting of these minds within a framework which still recognizes the desirability of a relatively free range for decision-making in the hands and mind of the individual entrepreneur, and which, at the same time, leaves the society assured that it will not be subject to breakdowns and discontinuities as it proceeds along the path of economic development and the fulfillment of its international responsibilities.

BIBLIOGRAPHY

ABRAMOWITZ, M., "Economics of Growth," in *A Survey of Contemporary Economics,* ed. B. F. Haley, Homewood, Illinois, Richard D. Irwin, 1952.

ADAMS, W., "A Critical Evaluation of Public Regulation by Independent Commissions: The Role of Competition in the Regulated Industries," *American Economic Review, 48* (1958), 527–43.

ALCHIAN, A. A., "Uncertainty, Evolution, and Economic Theory," *Journal of Political Economy, 58* (1950), 211–21.

ALDERFER, E. B. and MICHL, H. E., *Economics of American Industry,* 3d. ed., New York, McGraw-Hill, 1957.

ALEXANDER, S. S., "Issues of Business Cycle Theory Raised by Mr. Hicks," *American Economic Review, 41* (1951), 861–78.

AMERICAN IRON AND STEEL INSTITUTE, *Annual Statistical Report,* various years.

———, "Steel Capacity Statistics Based on Performance," *Steel Facts, 113,* April, 1952.

———, *Steelways,* January, 1950.

———, *Yearbook of the Iron and Steel Institute,* various years.

ANDREWS, P. W. S. and BRUNNER, E., *Capital Development in Steel: A Study of the United Steel Companies, Ltd.,* New York, Augustus M. Kelley, 1951.

ARMCO STEEL CORPORATION, *Annual Reports,* various years.

ARROW, K. J. and HOFFENBERG, M., (with the assistance of H. Markowitz and R. Shephard), *A Time Series Analysis of Inter-Industry Demand,* Amsterdam, North Holland Publishing Co., 1959.

AYRES, L. P., *Business Bulletin,* Cleveland Trust Co., Cleveland, Ohio, December 15, 1925.

BAKKE, E. W., *The Fusion Process,* New Haven, Labor and Management Center, Yale University, 1953.

BARLOON, M., "The Question of Steel Capacity," *Harvard Business Review, 27* (1949), 209–36.

BAUMOL, W. J., *Business Behavior, Value, and Growth,* New York, Macmillan, 1959.

———, *Economic Dynamics,* New York, Macmillan, 1951.

———, "Proposals for Increasing the Growth of National Output," in *What Price Economic Growth?,* ed. K. Knorr and W. J. Baumol, Englewood Cliffs, New Jersey, Prentice Hall, 1961.

BENDIX, R., "Bureaucracy and the Problem of Power," in *A Reader in Bureaucracy,* ed. R. K. Merton, A. P. Gray, B. Hickey, and H. C. Selvin, Glencoe, Illinois, The Free Press, 1952.

BENNER, S., "Benner's Prophecies of Future Ups and Downs in Prices, 1884," *Steelways,* January, 1950.

BERELSON, B., *Content Analysis in Communication Research,* Glencoe, Illinois, The Free Press, 1952.

BERNSTEIN, M. H., *Regulating Business by Independent Commission,* Princeton, Princeton University Press, 1955.

BERNSTEIN, P. L., "Profit-Theory—Where Do We Go From Here?" *The Quarterly Journal of Economics, 67* (1953), 407–22.

BETHLEHEM STEEL COMPANY, INC., *Bethlehem Review, 75,* January, 1959.

BLOUGH, R. M., *Free Man and the Corporation,* New York, McGraw-Hill, 1959.

———, "Learning to Multiply and Divide," an Address before the Economic Club of New York, January 15, 1957, New York, U.S. Steel Corporation, 1957.

———, "Steel in Perspective," a Talk before the Newcomen Society in North America, New York, December 4, 1958, New York, U.S. Steel Corporation, 1958.

BOSCHAN, P., "Productive Capacity, Industrial Production and Steel Requirements," in *Long Range Economic Projection,* a Report of the National Bureau of Economic Research, Princeton, Princeton University Press, 1954.

BOULDING, K., *The Organizational Revolution: A Study in the Ethics of Economic Organization,* New York, Harper, 1953.

———, "Welfare Economics," in *A Survey of Contemporary Economics, 2,* ed. B. F. Haley, Homewood, Illinois, Richard D. Irwin, 1952.

BOWLES, C., "The Prospects for a Connecticut Steel Mill: A Report to the People of Connecticut," Hartford, February, 1950.

BROUDE, H. W., "Locational Decisions and Their Relation to Industrial Development," *Explorations in Entrepreneurial History, 3* (1950), 103–34.

———, "The Role of the State in American Economic Development,

1820–1890," in *The State and Economic Growth,* ed. H. G. J. Aitken, New York, Social Science Research Council, 1959.

BUCHANAN, N. S., "Anticipations and Industrial Investment Decisions," *American Economic Review, 32,* supplement (1942), 141–55.

BURN, D. L., "Recent Trends in the History of the Steel Industry," *The Economic History Review, 17* (1947), 95–102.

BURNS, A. F. and MITCHELL, W. C., *Measuring Business Cycles,* New York, National Bureau of Economic Research, 1947.

CHENERY, H. B., "Overcapacity and the Acceleration Principle," *Econometrica, 20* (1952), 1–26.

CHRIST, C., "A Test of an Econometric Model for the United States, 1921–1947," in *Conference on Business Cycles,* New York, National Bureau of Economic Research, 1951.

CLARK, C., "A System of Equations Explaining the United States Trade Cycle, 1921–1941," *Econometrica, 17* (1949), 93–124.

CLARK, J. M., *Preface to Social Economics,* New York, Farrar & Rinehart, 1936.

———, *Strategic Factors in Business Cycles,* New York, National Bureau of Economic Research, 1935.

CLARK, P. G., "The Telephone Industry: A Study in Private Investment," in *Studies in the Structure of the American Economy,* ed. W. Leontief, New York, Oxford University Press, 1953.

CLARK, V. S., *History of Manufactures in the United States,* New York, McGraw-Hill, 1929.

CLEMENCE, R. V., ed., *Essays of J. A. Schumpeter,* Cambridge, Massachusetts, Addison-Wesley Press, 1951.

COLLINS, E. H., "Facts on Steel Production," *The New York Times,* October 23, 1950.

CORNFIELD, J., Evans, W. D., and Hoffenberg, M., "Full Employment Patterns 1950," *Monthly Labor Review, 64* (1947), 163–90.

COVERDALE AND COLPITTS, CONSULTING ENGINEERS, "Report on Proposed New England Steel Mill for New England Steel Development Corporation," and "Letter of Transmittal," April 17, 1951, processed.

CUMBERLAND, J. H., "The Locational Structure of East Coast Steel Industry," unpublished Ph.D. dissertation, Department of Economics, Harvard University, 1951.

DAUGHERTY, C.M., DE CHAZEAU, M. G., and STRATTON, S. S., *The Economics of the Iron and Steel Industry,* New York, McGraw-Hill, 1937.

DEAN, J., *Managerial Economics,* New York, Prentice-Hall, 1951.

DE CHAZEAU, M. G., "Regularization of Fixed Capital Investment by the Individual Firm," in *Regularization of Business Investment,* a report of the National Bureau of Economic Research, Princeton, Princeton University Press, 1954.

———, *The Adequacy of Steel Ingot Capacity,* Office of Production Management, Washington, December 10, 1940.

DOBROVOLSKY, S. P., *Corporate Income Retention, 1915–43,* New York, National Bureau of Economic Research, 1951.

DOMAR, E. D., "Capital Expansion, Rate of Growth and Employment," *Econometrica, 14* (1946), 137–47.

———, "Expansion and Employment," *American Economic Review, 37* (1947), 34–55.

———, "The Problem of Capital Accumulation," *American Economic Review, 38* (1948), 777–94.

DUESENBERRY, J. S., *Business Cycles and Economic Growth,* New York, McGraw-Hill, 1958.

———, "Hicks on the Trade Cycle," *The Quarterly Journal of Economics, 64* (1950), 464–76.

DUNN, G., "Report to the President of the United States on the Adequacy of the Steel Industry for National Defense," Office of Production Management, Washington, February 22, 1941.

———, "Second Report to the President of the United States on the Adequacy of the Steel Industry for National Defense," Office of Production Management, Washington, May 22, 1941.

ECKAUS, R. S., "The Acceleration Principle Reconsidered," *The Quarterly Journal of Economics, 67* (1953), 209–30.

ECKSTEIN, O. and FROMM, G., "Steel and the Postwar Inflation," Study Paper No. 2, prepared in connection with the Study of Employment, Growth, and Price Levels for consideration by the Joint Economic Committee, Congress of the United States, November 6, 1959, 86th Congress, 1st Session, Washington, Government Printing Office, 1959.

EELLS, R., *The Meaning of Modern Business,* New York, Columbia University Press, 1960.

EISNER, R., "Guaranteed Growth of Income," *Econometrica, 21* (1953), 169–71.

———, "Interview and Other Survey Techniques and the Study of Investment," Conference on Research in Income and Wealth, New York, National Bureau of Economic Research, 1953. This study was later revised and published as *Determinants of Capital Expenditures: An Interview Study,* Studies in Business Expectations and Planning, Urbana, University of Illinois, 1956.

EVANS, G. C., "The Dynamics of Monopoly," *American Mathematical Monthly, 31* (1924), 77–83.

EVANS, W. D. and HOFFENBERG, M., "The Interindustry Relations Study for 1947," *Review of Economics and Statistics, 34* (1952), 97–142.

FAIRLESS, B. F., *It Could Happen Only in the U.S.*, New York, Time, Inc., 1956.

FELLNER, W., "Employment Theory and Business Cycles," in *A Survey of Contemporary Economics*, ed. H. S. Ellis, Philadelphia, Blakiston, 1948.

———, "Long-Term Projections of Capital Formation," Conference on Research in Income and Wealth, New York, National Bureau of Economic Research, 1951.

———, *Monetary Policies and Full Employment*, Berkeley, University of California Press, 1947.

———, "The Technological Argument of the Stagnation Thesis," *The Quarterly Journal of Economics, 55* (1949), 638–51.

FELS, R., "The Theory of Business Cycles," *The Quarterly Journal of Economics, 66* (1952), 25–42.

FLORENCE, P. S., *Investment Location, and Size of Plant*, Cambridge, England, Cambridge University Press, 1948.

FOGARTY, F., "Needed Soon: More Materials Capacity," *Architectural Forum, 109* (November, 1958), 140–41.

Fortune, "U.S. Steel," March, April, June, 1936.

FRIENDS OF DOMESTIC INDUSTRY, "On the Product and Manufacture of Iron and Steel," Reports of Committees of The General Convention of the Friends of Domestic Industry assembled at New York, October 26, 1831.

FRISCH, R., "The Interrelation Between Capital Production and Consumer-taking," *Journal of Political Economy, 39* (1931), 646–54.

GLASER, E., "The Interindustry Economics Research Program of the Federal Government," Conference on Research in Income and Wealth, New York, National Bureau of Economic Research, 1952.

GOODWIN, R. M., "Econometrics in Business-Cycle Analysis," in *Business Cycles and National Income*, by A. H. Hansen, New York, W. W. Norton, 1951.

———, *Proceedings of a Conference on Inter-Industrial Relations*, Driebergen, Holland, 1958.

———, "The Nonlinear Accelerator and the Persistence of Business Cycles," *Econometrica, 19* (1951), 1–17.

GORDON, R. A., *Business Leadership in the Large Corporation*, Washington, The Brookings Institution, 1945.

———, "Investment Opportunities in the United States before and after World War II," in *The Business Cycle in the Post-War World*, ed. E. Lundberg, London, Macmillan, 1955.

GRESS, H. L. and YNTEMA, T. O., "A Statistical Analysis of the Demand for Steel, 1919–1938," Temporary National Economic Committee Paper, pamphlet No. 5, United States Steel Corporation.

GROSSE, A., "Textile Production Functions, Equipment Requirements and Technological Change—A Study in the Utilization of Primary Technical Information," Cambridge, Massachusetts, Harvard Economic Research Project, December, 1946.

GROSSE, A. and DUESENBERRY, J. S., "Technological Change and Dynamic Models," Conference on Research in Income and Wealth, New York, National Bureau of Economic Research, 1952.

GROSSE, R. N. and BERMAN, E. B., "Estimating of Future Purchases of Capital Equipment for Replacement," in *Problems of Capital Formation,* a report of the Conference on Research in Income and Wealth, National Bureau of Economic Research, Princeton, Princeton University Press, 1957.

HABERLER, G., *Prosperity and Depression,* 3d ed., New York, United Nations, 1946.

HAMBERG, D., *Economic Growth and Instability: A Study in the Problem of Capital Accumulation, Employment, and the Business Cycle,* New York, W. W. Norton, 1956.

HANSEN, A. H., *Business Cycles and National Income,* New York, W. W. Norton, 1951.

HARROD, R. F., "An Essay in Dynamic Theory," *Economic Journal, 49* (1939), 14–33.

———, *Economic Essays,* New York, Harcourt Brace, 1952.

———, *Towards a Dynamic Economics,* London, Macmillan, 1948.

HART, A. G., *Anticipations, Uncertainty and Dynamic Planning,* New York, Augustus M. Kelley, 1951.

HAUCK, W. A., "Status of the Steel Expansion Program," War Production Board, Washington, June 30, 1942.

———, "Steel Expansion Program," submitted to Office of Production Management, Washington, September 24, 1941.

HELLER, W., "The Anatomy of Investment Decisions," *Harvard Business Review, 29* (1951), 95–103.

HENRY, W. E., "The Business Executive: The Psychodynamics of a Social Role," *The American Journal of Sociology, 54* (1949), 286–91.

HICKMAN, B. G., "Capacity, Capacity Utilization, and the Acceleration Principle," in *Problems of Capital Formation,* a report of the Conference on Research in Income and Wealth, National Bureau of Economic Research, Princeton University Press, 1957.

HICKS, J. R., *A Contribution to the Theory of the Trade Cycle,* London, Oxford University Press, 1950.

———, *Value and Capital,* Oxford, Clarendon Press, 1946.

INLAND STEEL COMPANY, *Annual Report,* 1948.
INTERNATIONAL LABOR ORGANIZATION, IRON AND STEEL COMMITTEE, *Regularization of Production and Employment at a High Level,* report no. 2, second session, Stockholm, 1947.
IRON AND STEEL BOARD, *Development in the Iron and Steel Industry, Special Report, 1961,* London, Her Majesty's Stationery Office, 1961.
ISARD, W. and CUMBERLAND, J. H., "New England as a Possible Location for an Integrated Iron and Steel Works," *Economic Geography, 26* (1950), 245–59.
——— and KUENNE, R. W., "The Impact of Steel Upon the Greater New York-Philadelphia Industrial Region: A Study in Agglomeration Projection," *Review of Economics and Statistics, 35* (1953), 289–301.

JONES, R. C., "The Effects of Inflation on Capital and Profit: The Record of Nine Steel Companies," *The Journal of Accountancy, 87,* (1949).

KALDOR, N., "A Model of the Trade Cycle," *Economic Journal, 50* (1940), 78–92.
KAPLAN, A. D. H., DIRLAM, J. B., and LANZILLOTTI, R. F., *Pricing and Big Business: A Case Approach,* Washington, The Brookings Institution, 1958.
KAYSEN, C., "A Dynamic Aspect of the Monopoly Problem," *Review of Economics and Statistics, 31* (1949), 109–13.
———, "Basing Point Pricing and Public Policy," *The Quarterly Journal of Economics, 42* (1949), 289–314.
KEYNES, J. M., *The General Theory of Employment, Interest and Money,* New York, Harcourt Brace, 1936.
KLEIN, L., *Economic Fluctuations in the United States,* New York, John Wiley, 1950.
———, "Reply," to Leontief's "Comment on 'Studies in Investment Behavior' " in *Conference on Business Cycles,* New York, National Bureau of Economic Research, 1951.
———, "Studies in Investment Behavior," in *Conference on Business Cycles,* New York, National Bureau of Economic Research, 1951.
KNAUTH, O., *Managerial Enterprise: Its Growth and Methods of Operation,* New York, W. W. Norton, 1948.
KNIGHT, F. H., *The Ethics of Competition,* New York, Augustus M. Kelley, 1951.
KNOX, A. D., "The Acceleration Principle and the Theory of Investment," *Economica,* n.s. *19* (1952), 269–97.
KOOPMANS, T. C., "Comment" on W. Leontief, "Input-Output Analysis and its Use in Peace and War Economies: Recent Developments in

the Study of Interindustrial Relationships," *American Economic Review, 34* (1949), 234.

KRUG, J. A., *National Resources and Foreign Aid,* U.S. Department of Commerce, Washington, Government Printing Office, 1947.

KUH, E. and MEYER, J. R., "Correlation and Regression Estimates when the Data Are Ratios," *Econometrica, 23* (1955), 400–16.

KUZNETS, C., *National Income and Capital Formation, 1919–1935,* New York, National Bureau of Economic Research, 1937.

———, *National Income and Its Composition, 1919–1938,* 2 vols., New York, National Bureau of Economic Research, 1941.

———, *Secular Movements in Production and Prices,* Boston, Houghton Mifflin, 1930.

———, "Static and Dynamic Economics," *American Economic Review, 20* (1930), 426–41.

———, "The Relation between Capital Goods and Finished Products in the Business Cycle," *Economic Essays in Honor of Wesley Clair Mitchell,* New York, Columbia University Press, 1935.

LANE, R. E., *The Regulation of Businessmen: Social Conditions of Government Economic Control,* New Haven, Yale University Press, 1954.

LEONTIEF, W., "Comment on 'Studies in Investment Behavior,'" in *Conference on Business Cycles,* New York, National Bureau of Economic Research, 1951.

———, "Computational Problems Arising in Connection with Economic Analysis of Industrial Relationships," *Proceedings of a Symposium on Large-Scale Digital Calculating Machinery,* Cambridge, Massachusetts, Harvard University Press, 1948.

———, "Interrelations of Prices, Output, Savings, and Investment," *Review of Economic Statistics, 19* (1937), 109–32.

———, "Output, Employment, Consumption and Investment," *The Quarterly Journal of Economics, 58* (1944), 290–314.

———, "Postulates: Keynes *General Theory* and the Classicists," in *The New Economics,* ed. S. E. Harris, New York, Alfred A. Knopf, 1948.

———, *Progress Report, Preliminary,* Cambridge, Massachusetts, Harvard Economic Research Project, December, 1952, processed.

———, "Quantitative Input and Output Relations in the Economic System of the United States," *Review of Economic Statistics, 18* (1936), 105–25.

———, "Recent Developments in the Study of Interindustry Relations," *American Economic Review, 39* (1949), 211–25.

———, "Some Basic Problems of Empirical Input-Output," Conference on Research in Income and Wealth, National Bureau of Economic Research, 1952.

———, *Studies in the Structure of the American Economy,* New York, Oxford University Press, 1953.

———, "The Economics of Industrial Interdependence," *Dun's Review,* February, 1946.

———, *The Structure of American Economy, 1919–1939,* 2d ed., revised, New York, Oxford University Press, 1951.

——— and Isard, W., "The Extension of Input-Output Techniques to Interregional Analysis," *Studies in the Structure of the American Economy,* New York, Oxford University Press, 1953.

LEWIS, B. W., "Economics by Admonition," *American Economic Review, 49* (1959), 384–98.

LISTER, L., *Europe's Coal and Steel Community: An Experiment in Economic Union,* New York, Twentieth Century Fund, 1961.

LUTZ, F. and LUTZ, V., *The Theory of Investment of the Firm,* Princeton, Princeton University Press, 1951.

MACCULLUM, E.C., *The Iron and Steel Industry in the United States,* London, King, 1931.

MACDONALD, J., "How Executives Make Decisions," in *The Executive Life,* by the Editors of *Fortune,* New York, Doubleday, 1956.

MACHLUP, F., *The Basing Point System,* Philadelphia, Blakiston, 1949.

MACK, R. P., *The Flow of Business Funds and Consumer Purchasing Power,* New York, Columbia University Press, 1941.

MAURER, H., *Great Enterprise: Growth and Behavior of the Big Corporation,* New York, Macmillan, 1955.

MCGRAW-HILL DEPARTMENT OF ECONOMICS, "Business Plans for New Plants and Equipment, 1947–1960," Tenth Annual Survey, New York, McGraw-Hill, 1960.

MEYER, J. R. and KUH, E., *The Investment Decision: An Empirical Study,* Cambridge, Mass., Harvard University Press, 1957.

MILLER, J. P., "Contribution of Industrial and Human Relations Research to Economists' Theory of the Firm," Proceedings of the Eleventh Annual Meeting of the Industrial Relations Research Association, *11* (1958), 134–42.

———, "The Pricing Effects of Accelerated Amortization," *Review of Economics and Statistics, 34* (1952), 10–17.

MODIGLIANI, F., "Comment on 'Capacity, Capacity Utilization, and the Acceleration Principle,' " in *Problems of Capital Formation,* a report of the Conference on Research in Income and Wealth, New York, National Bureau of Economic Research, 1957.

MOL, A. A., "Does the Nation Need More Steel Capacity? Projection of

Past Trends Suggests Sharp Production Decline in Years Ahead," *Barron's Weekly,* January 17, 1949.

NATIONAL ASSOCIATION OF MANUFACTURERS, *The Iron and Steel Industry in the Defense Economy,* New York, National Association of Manufacturers, 1941.

NATIONAL BUREAU OF ECONOMIC RESEARCH, *Regularization of Business Investment,* Princeton, Princeton University Press, 1954.

———, *Short Term Economic Forecasting,* Princeton, Princeton University Press, 1955.

NETREBA, S. S., "The Development of the Bill of Goods for Interindustry Analysis," Conference on Research in Income and Wealth, New York, National Bureau of Economic Research, 1952.

NEWCOMER, M., *The Big Business Executive: The Factors that Made Him,* New York, Columbia University Press, 1955.

OKUN, A. M., "The Value of Anticipations Data in Forecasting National Product," in *Quality and Economic Significance of Anticipations Data,* National Bureau of Economic Research, special conference series, *10,* Princeton, Princeton University Press, 1960.

PAPANDREOU, A. G., "Some Basic Problems in the Theory of the Firm," in *A Survey of Contemporary Economics, 2,* ed. B. F. Haley, Homewood, Illinois, Richard D. Irwin, 1952.

PARRISH, J. B., "Iron and Steel in the Balance of World Power," *Journal of Political Economy, 64* (1956), 369–88.

PERSONS, W. M., "The Iron and Steel Industry during Business Cycles," *Review of Economic Statistics, 3* (1921), 378–83.

PHILLIPS, A., "Plant Capacity: A Study in the Non-Friction Bearing Industry," unpublished Ph.D. dissertation, Department of Economics, Harvard University, 1953.

PRESIDENT'S MATERIALS POLICY COMMISSION, *The Outlook for Key Commodities, Resources for Freedom, 2,* A Report to the President, Washington, Government Printing Office, 1952.

REPUBLIC STEEL CORPORATION, *Annual Reports,* Cleveland, various years.

RICH, J. L., "Techniques and Uses of Forecasts of General Business Conditions in U.S. Steel," *Proceedings of the Business and Economics Section, American Statistical Association,* Washington, 1956.

ROBERTSON, D. H., "A Revolutionist's Handbook," *The Quarterly Journal of Economics, 64* (1950), 1–14.

———, "Thoughts on Meeting Some Important Persons," *The Quarterly Journal of Economics, 68* (1954), 181–90.

ROCKEFELLER BROTHERS FUND, *The Challenge to America: Its Economic and Social Aspects,* New York, Doubleday, 1958.

ROOS, C. F., "Dynamic Theory of Economics," *Journal of Political Economy, 35* (1929), 632–56.

SAMUELSON, P. A., "Dynamic Process Analysis," in *A Survey of Contemporary Economics,* ed. H. S. Ellis, Philadelphia, Blakiston, 1948.

———, "Dynamics, Statics, and the Stationary States," *Review of Economic Statistics, 25* (1943), 58–68.

———, "The Interactions Between the Multiplier Analysis and the Principle of Acceleration," in *Readings in Business Cycle Theory,* Philadelphia, Blakiston, 1944.

SCHELLING, T. C., "Capital Growth and Equilibrium," *American Economic Review, 37* (1947), 864–76.

SCHROEDER, G. G., *The Growth of Major Steel Companies, 1900–1950,* Baltimore, Johns Hopkins Press, 1953.

SCHUMPETER, J. A., "Review of Keynes' *General Theory," Journal of the American Statistical Association, 31* (1936), 791–95.

———, *The Theory of Economic Development,* Cambridge, Mass., Harvard University Press, 1949.

SELZNIK, P., "Coöptation: A Mechanism for Organizational Stability," in *Reader in Bureaucracy,* ed. R. K. Merton, A. P. Gray, B. Hickey, and H. C. Selvin, Glencoe, Illinois, The Free Press, 1952.

———, "A Theory of Organizational Commitments," in *Reader in Bureaucracy,* ed. R. K. Merton, A. P. Gray, B. Hickey, and H. C. Selvin, Glencoe, Illinois, The Free Press, 1952.

SIMON, H. A., *Administrative Behavior: A Study of Decision-Making Processes in Administrative Organization,* New York, Macmillan, 1940.

———, "Theories of Decision-Capital Making in Economics and Behavioral Sciences," *American Economic Review, 49* (1959), 253–83.

SIMS, C. E., "The Technology of Iron and Steel," in *The Promise of Technology, Resources for Freedom, 4,* a report to the President by the President's Materials Policy Commission, Washington, Government Printing Office, 1952.

SMITH, C. A., "The Cost-Output Relation for the United States Steel Corporation," *Review of Economic Statistics, 24* (1942), 166–76.

SMITHIES, A., "Aspects of the Basing-Point System," *American Economic Review, 32* (1942), 705–26.

STAEHLE, H., "The Measurement of Statistical Cost Functions: An Appraisal of Some Recent Contributions," *American Economic Review, 32* (1942), 321–33.

STANDARD AND POOR'S *Industry Surveys,* June 27, 1947.

Steel, "Get Ready for the New Boom," November 17, 1958.

STEINDL, J., *Maturity and Stagnation in American Capitalism,* Oxford, Blackwell, 1952.

STELZER, I. M., "Review of *Market Power: Size and Shape Under the Sherman Act* by G. E. Hale and R. D. Hale," *The American Economic Review, 49* (1959), 1110.

STIGLER, G., *Production and Distribution Theories,* New York, Macmillan, 1949.

STRIKE, C. S., Chairman, Connecticut Steel Advisory Committee, Letter to Honorable John Davis Lodge, Governor of the State of Connecticut, Hartford, April 20, 1951.

TERBORGH, G., *Dynamic Equipment Policy,* New York, McGraw-Hill, 1949.

TINBERGEN, J., *Business Cycles in the United Kingdom, 1870–1914,* Amsterdam, North Holland Publishing Co., 1951.

———, *Statistical Testing of Business Cycle Theories, I, A Method and its Application to Investment Activity,* Geneva, League of Nations Economic Intelligence Service, 1939.

———, *Statistical Testing of Business Cycle Theories, II, Business Cycles in the United States, 1919–1932,* Geneva, League of Nations Economic Intelligence Service, 1939.

TOBIN, J., "The Business Cycle in the Post-War World: A Review," *The Quarterly Journal of Economics, 73* (1958), 284–91.

TOWER, W. S., "Steel in a Year of War," an Address to the 51st General Meeting of the American Iron and Steel Institute, New York, May 21, 1942.

———, Address to the 53d General Meeting of the American Iron and Steel Institute, New York, May 25, 1944.

TRUMAN, H. S., "Economic Message to Congress," Washington, Government Printing Office, January 12, 1951.

———, "State of the Union Message to the Congress of the United States," Washington, Government Printing Office, January 5, 1949.

TSIANG, S. C., "Accelerator, Theory of the Firm and the Business Cycle," *The Quarterly Journal of Economics, 65* (1951), 325–41.

———, "Rehabilitation of Time Dimension of Investment in Macrodynamic Analysis," *Economica, n.s. 16* (1949), 204–17.

U.S. BUREAU OF THE CENSUS, *Historical Statistics of the United States, 1789–1945,* Washington, Government Printing Office.

U.S. CONGRESS, House of Representatives. *Hearings before the Subcommittee on Study of Monopoly Power of the Committee on the*

Judiciary, 81st Congress, 2d Session, Washington, Government Printing Office, 1950.

———, *Hearings,* and *Staff Report on Employment, Growth, and Price Levels before the Joint Economic Committee,* 86th Congress, 1st Session.

———, Senate, *Final Report of the Special Committee to Study Problems of American Small Business,* 80th Congress, 2d Session, Washington, Government Printing Office, 1949.

———, *Hearing before the Special Committee to Study Problems of American Small Business,* part 8, "Steel Supply and Distribution Problems Affecting Smaller Manufacturers and Users," 80th Congress, Washington, Government Printing Office, 1947.

———, *Interim Report of the Special Committee to Study Problems of American Small Business,* 80th Congress, 2d Session, Washington, Government Printing Office, 1948.

———, "War Plant Disposal—Iron and Steel Plants," *Joint Hearing before Sub-Committee on Surplus Property of the Committee on Military Affairs and the Industrial Reorganization Sub-Committee of the Special Committee on Post-War Economic Policy and Planning,* 79th Congress, 1st Session, November, 1945, Washington, Government Printing Office, 1946.

———, *Temporary National Economic Committee Hearings,* 76th Congress, 3d Session, parts 19 and 26, November 6–10, 1939 and January 23–25, 1940, Washington, Government Printing Office, 1940.

———, "Investments, Profits, and Rate of Return for Selected Industries," *Study submitted by Federal Trade Commission to Temporary National Economic Committee,* 76th Congress, 3d Session, part 31, Washington, Government Printing Office, 1941.

U.S. Department of Commerce, *Business Statistics, 1951,* Washington, Government Printing Office, 1951.

———, Office of Business Economics, "Investment Cost and Capacity in Iron and Steel: An Exploratory Study," prepared by the Industrial Capacity Section, Business Structure Division for the Interindustry Economic Research Program of the U.S. Department of the Air Force, September, 1953 (processed).

———, Office of Business Economics, "Steel Ingots and Finished Steel: Flow Inputs and Outputs," Industrial Capacity Section under general supervision of Murray Foss (processed).

———, *The U.S. Industrial Outlook for 1961,* Washington, Government Printing Office, 1961.

U.S. Department of Labor, Bureau of Labor Statistics, *Monthly Labor Review, 64,* February, March, 1947.

U.S. National Recovery Administration, "Code of Fair Competition for the Iron and Steel Industry," Article V, Section 2, Washington, Government Printing Office.

U.S. NATIONAL RESOURCES COMMITTEE, *Capital Requirements, A Study in Methods as Applied to the Iron and Steel Industry,* Washington, Government Printing Office, 1940.

———, *Patterns of Resource Use,* Washington, Government Printing Office, 1938.

U.S. News and World Report, December 1, 1950, 38.

VANDERBLUE, H. B. and CRUM, W. L., *The Iron Industry in Prosperity and Depression,* Chicago, A. W. Shaw, 1927.

WEIR, E. T., "Steel Expansion: The Problems Involved," an Address before the New York Society of Security Analysts, New York, March 22, 1956.

WELSH, E. C., "Government Aid to Business Expansion," *American Economic Review, 42* (1952), 418–27.

WHITMAN, R. W., "The Statistical Law of Demand for a Producers' Good as Illustrated by the Demand for Steel," *Econometrica, 4* (1936), 138–52.

WILEY, K. H., and EZECKIEL, M., "The Cost Curve for Steel Production," *Journal of Political Economy, 48* (1940), 777–821.

ZABEL, E., *Concepts and Measurement of Productive Capacity,* Technical Report issued under Office of Naval Research Contract, Princeton University, November 1955.

INDEX

YALE STUDIES IN ECONOMICS

1. Richard B. Tennant. The American Cigarette Industry. A Study in Economic Analysis and Public Policy
2. Kenneth D. Roose. The Economics of Recession and Revival. An Interpretation of 1937–38
3. Allan M. Cartter. The Redistribution of Income in Postwar Britain. A Study of the Effects of the Central Government Fiscal Program in 1948–49
4. Morton S. Baratz. The Union and the Coal Industry
5. Henry C. Wallich. Mainsprings of the German Revival
6. Lloyd G. Reynolds and Cynthia H. Taft. The Evolution of Wage Structure
7. Robert Triffin. Europe and the Money Muddle
8. Mark Blaug. Ricardian Economics. A Historical Study
9. Thomas F. Dernburg, Richard N. Rosett, and Harold W. Watts. Studies in Household Economic Behavior
10. Albert O. Hirschman. The Strategy of Economic Development
11. Bela A. Balassa. The Hungarian Experience in Economic Planning.
12. Walter C. Neale. Economic Change in Rural India
13. John M. Montias. Central Planning in Poland
14. Paul W. MacAvoy. Price Formation in Natural Gas Fields. A Study of Competition, Monopsony, and Regulation
15. Harry A. Miskimin. Money, Prices, and Foreign Exchange in Fourteenth-Century France
16. Henry W. Broude. Steel Decisions and the National Economy
17. Robert M. Macdonald. Collective Bargaining in the Automobile Industry. A Study of Wage Structure and Competitive Relations.